H. Haken A. Mikhailov (Eds.)

Interdisciplinary Approaches to Nonlinear Complex Systems

With 92 Figures

Springer-Verlag
Berlin Heidelberg New York
London Paris Tokyo
Hong Kong Barcelona
Budapest

Professor Dr. Dr. h. c. Hermann Haken

Institut für Theoretische Physik und Synergetik der Universität Stuttgart,
Pfaffenwaldring 57/IV, 70569 Stuttgart 80, Fed. Rep. of Germany and
Center for Complex Systems, Florida Atlantic University,
Boca Raton, FL 33431, USA

Professor A. Mikhailov

Fritz-Haber-Institut der Max-Planck-Gesellschaft,
Faradayweg 4–6, 14195 Berlin 33

Series Editor:

Professor Dr. Dr. h. c. Hermann Haken

Institut für Theoretische Physik und Synergetik der Universität Stuttgart,
Pfaffenwaldring 57/IV, 70569 Stuttgart 80, Fed. Rep. of Germany and
Center for Complex Systems, Florida Atlantic University,
Boca Raton, FL 33431, USA

ISBN 978-3-642-51032-8 ISBN 978-3-642-51030-4 (eBook)
DOI 10.1007/978-3-642-51030-4

Typesetting: Camera ready by authors

57/3140-5 4 3 2 1 0 – Printed on acid-free paper

Preface

A wide range of scientific disciplines, including biology, ecology, psychology, cognitive science, economics and sociology, involve the study of complex systems, in which a decisive role is played by the nonlinear interactions between many elements. A central aim is to explain how such interactions can bring about qualitatively new structures which determine the behaviour of the entire system and which are not reducible to a sum of the individual effects.

The emergence of coherent cooperative behaviour is the central theme of this book. Despite the great diversity in the nature of possible complex systems they all rely on essentially the same basic principles of autonomous organization and hierarchical control. Consequently, the mathematical models employed to describe them tend to be similar in all cases.

This common ground provides an opportunity for fruitful interdisciplinary contacts leading to more intensive exchange of concepts and models between different fields. To make such contacts successful, the terminological straitjackets corresponding to the details of specific applications must be carefully removed and the essence of the phenomena involved needs to be made clear to outsiders without significant expertise in the given narrow field.

The contributions to this volume are based on the talks presented at the Workshop "Interdisciplinary Approaches to Nonlinear Complex Systems" which took place at the Centre for Interdisciplinary Studies in Bielefeld, Germany, on 19–23 October 1992. A unique feature of this event was that it hosted a group of actively working scientists from very diverse research directions who aimed to explain to one another (and now also to the readers of this book) the current situation and the research outlook in their respective disciplines from the viewpoint of the theory of complex systems. The review lectures were followed by long discussions and lively exchanges of opinions between the participants. In addition to the material included in this volume, talks were also given by H. P. Koepchen, D. Lehmann, H. Meinhardt and V. Sergeev.

We want to express our gratitude to the Centre for Interdisciplinary Studies of Bielefeld University and its director Prof. P. Weingart for the administrative assistance and financial support in organizing this meeting. We would especially like to thank Ms. T. Valentin who has helped us so much, both with the preparation and during the course of the Workshop.

Stuttgart, Berlin
May 1993

H. Haken
A. Mikhailov

Contents

Introduction

H. Haken and A. Mikhailov

For a very long time scientific discourse has been dominated by the reductionist paradigm. This saw the goal of science as being the reduction of observed phenomena to the elementary entities occupying the lower levels of the material hierarchy. Hence, an explanation of all biological processes was ultimately sought in terms of chemistry (as is indeed clear in the field of molecular biology) while psychological phenomena were generally attributed to some underlying physiological changes. A consequence of such a paradigm has been a special attention to the contacts between neighbouring disciplines, with a tacit assumption that a science dealing with more fundamental (simpler) entities occupies the superior position. Only basic sciences such as physics or chemistry were expected to have clear mathematical constructions, while disciplines dealing with more complex systems remained essentially descriptive.

Although reductionism brings logical order into the scientific edifice it does not provide true integrity to the realm of science. It yields no common ground for the contacts between distant disciplines and does not stimulate such contacts. Today, however we are witnessing a rapidly growing interest in interdisciplinary communication, especially between different branches of the life sciences. What might be the origins of this tendency?

There is accumulating evidence that certain motifs and scenarios are frequently repeated in all fields of the life sciences (resembling to some extent the recurrence of myths and archetypes in different human cultures). These findings, which can no longer be explained by the material similarity of the elements involved, indicate deep analogies in the manner of organization of various living systems.

Although living systems may differ substantially in their detailed properties, they all share one common feature: They must function in a coherent and predictable way while being composed of a large number of different interacting units.

If we were to take an arbitrary aggregation of interacting elements, it would most probably demonstrate an irregular and unpredictable behaviour. It is known that great skill is required to design a complex system capable of purposeful operation. The natural systems studied in the life sciences are products of long evolutionary selection which has made their inner organization perfect.

There are several general principles of organization of complex systems. Firstly, such systems represent hierarchies. This means that they can be divided into different levels, each representing a subsystem which consists of relatively uniform elements that interact in a simple way with one another. These interactions are responsible for autonomous pattern formation at any given hierarchical level. The higher subsystems in the hierarchy provide control over the processes of pattern formation at the lower subordinate levels but do not directly interfere with these processes.

The detailed analysis of how these general principles are implemented in concrete biological or social systems is a task for the philosophy and methodology of science. But the unveiling of abstract universal principles may be of less interest for practically oriented scientists. By entering into interdisciplinary exchanges they hope to find a clue to their own research problems, to hear about models and approaches which might, after some modifications, be applicable to their own particular work.

Efficient communication requires the use of a common language. In the case of scientific discourse the elements of the language are basic concepts and relationships. Unfortunately, the independent and largely isolated development of different scientific disciplines has resulted in great diversity of terminology and very similar entities are often differently denoted in various fields. Therefore, an attempt needs to be made to bring closer at least the most essential terminology and especially that which deals with the aspects of internal organization and pattern formation in complex systems of various origins. This can only be achieved through intensive discussions between representatives of the different disciplines.

Talks that are successfully addressed to a multidisciplinary audience have a property which is reminiscent of works of art: While watching an antique tragedy one learns more than bare facts about life habits in ancient Greece. Its characters personify the universal archetypes of the human psyche. In a somewhat similar way, the "characters" of the interdisciplinary talks are powerful metaphors which are merely dressed up in the costume and language of a particular field.

Let us consider an example taken from organic chemistry. The Belousov-Zhabotinskii reaction became famous because of its rich potential for pattern formation, including the generation of travelling waves, pacemakers and spirals. The actual mechanism of this reaction is extremely complicated; it involves several tens of individual reaction stages and is not yet fully understood. Despite this, the system has become known far beyond the narrow field of organic chemistry and is frequently mentioned at interdisciplinary meetings. Whenever this is done, the details of the complex reaction scheme are usually left in the shadows while the pattern formation aspects are emphasized. For a multidisciplinary audience its significance is as an implementation of certain autonomous pattern formation mechanism rather than as a particular chemical phenomenon.

Mathematics is sometimes called the universal language of the sciences. This is probably not completely true since mathematics is really a science in its own right, with its own special tools and techniques. But it is definitely true to say that models formulated in mathematical terms can be much more readily communicated beyond the borders of the discipline of their origin. This explains the special interest in the mathematical models of cooperative phenomena evident at multidisciplinary meetings.

Having noted this, we must point out that not just any mathematical model can successfully enter the arena of interdisciplinary communication. We consider again a concrete example. Suppose that somebody has the task of predicting the behaviour of a complex ecological system which includes a hundred essential components. He or she then writes a hundred coupled differential equations which reflect the mass balance and the reproduction and other biological processes of the flora and fauna, takes further reasonable numerical values of the parameters involved and performs a numerical integration of all these equations. Such mathematical investigation may indeed be of much importance, yielding valuable practical predictions or suggesting a particular strategy for dealing with an ecological system However, this approach would be of very little interest from the viewpoint of interdisciplinary communication.

The mathematical models which play the role of linguistic elements should possess a generic property. They retain only the bare bones of the actually observed phenomena thereby sacrificing its minute details, even though the latter might be very important within a particular field. Therefore they could be described as abstractions or universal archetypes of the evolution and cooperative behaviour in complex systems of various origins.

The formulation of a generic model is a gradual process which involves the concerted efforts of representatives of different disciplines. But its first stage consists in proposing possible candidates based on studies of particular problems. This is yet another task of interdisciplinary meetings.

Keeping a stock of generic mathematical models is a service which could be best provided by the mathematicians. It should also include an analysis of the generic models and assistance in their adaptation to the needs of particular disciplines. Regrettably, traditional mathematics is not well suited for this. It sees its task in a logically rigorous and self-contained investigation of well-defined problems and at present pays almost no attention to how the models are actually applied in other fields of science.

It turns out that theoretical physicists probably have the most expertise in formulating and dealing with mathematical models. For a long time, part of their everyday routine has consisted in suggesting mathematical models that are simple but rich in content and which are able to grasp the most essential properties of experimentally observed phenomena. Moreover, theoretical physics also includes the special art of performing approximate semi-empirical calculations which do not obey the criteria of rigour set by the mathematicians but have nonetheless yielded very impressive results, which would have never been achieved if the mathematical protocol were strictly followed. This success relies on well-developed intuition and a good qualitative understanding of mathematical models.

Mathematical studies of complex systems do not constitute a separate scientific discipline. This subject has no linear construction proceeding from some fundamental axioms or basic data. Instead what we have here is merely a loose collection of generic mathematical models of cooperative phenomena that are abstractions derived from various fields of science. They possess only one common feature:

nonlinearity. Linear models cannot be applied to describe the emergent synergetic behaviour of an entire system, produced by the cooperation of its individual parts but not reducible to a sum of their effects. Indeed, the principal property of any linear equation is that any superposition of its solutions yields again a valid solution which thus indicates a lack of any cooperative effect.

In a sense, the internal structure discernable in the studies of complex systems can perhaps itself be best characterized as a kind of a complex system. Only when taken together, can the impressive baroque of linked models suddenly begin to resemble what it is intended to reproduce – life.

The contributors to this volume come from diverse fields of science, ranging from neurophysiology to the social and economic sciences. Despite the apparent mosaic of their contributions, all of them are united in that they represent efforts to formulate and to interpret the findings of particular disciplines in terms of complex systems. How well we have succeeded in this task is to be judged by our readers.

Synergetics as a Strategy to Cope with Complex Systems

H. Haken

Institute for Theoretical Physics and Synergetics,
Pfaffenwaldring 57/4, D-7000 Stuttgart 80 (Vaihingen),
Fed. Rep. of Germany

This contribution gives a brief outline of the three main approaches of synergetics: a) the microscopic approach based on evolution equations for the variables of the subsystems of a complex system b) the macroscopic approach based on the maximum information principle c) the phenomenological approach based on order parameter equations.

1 Introduction

Science is becoming more and more concerned with the study of complex systems. Complex systems are ubiquitous, biology abounds of them but also economy, ecology, sociology deal with such systems. These are systems that are composed of many parts that interact among each other in a complicated fashion. In a way, a general strategy of coping with such systems may be traced back to Décartes. According to him one has to decompose such a system into simpler and simpler parts until we can deal with these simple parts. This strategy is followed up in many disciplines. A prominent example is molecular biology. An important aspect has to be observed, however. By means of the cooperation of the individual parts of a system new properties may emerge that are not present at the level of the individual parts. In addition, in a number of systems it is not so easy - or even impossible - to identify simple individual components. Thus a search for other strategies is certainly necessary.

In the following I wish to give a brief outline of a strategy that is suggested by synergetics [1], [2].

2 Synergetics

As is well-known, synergetics is an interdisciplinary field of research that deals with systems that are composed of many individual parts and that may produce spatial, temporal, of functional structures by self- organization. In physics, chemistry, and biology the systems under consideration are open systems, i.e. they are driven by a constant influx of energy and/or matter. But synergetics deals with other systems, say in economy, ecology, and sociology as well. Quite generally speaking, the situations considered by synergetics are as follows: The

Springer Series in Synergetics, Vol. 62 **Interdisciplinary Approaches to Nonlinear Complex Systems** – Eds.: H. Haken and A. Mikhailov

systems are controlled from the outside in a rather unspecific manner by control parameters, e.g. by the energy flux into a system. When one or several control parameters are changed, the old state of the system can become destabilized and is then replaced by a new state which may exhibit specific kinds of spatial, temporal, or functional structures. The strategy of synergetics consists in considering those situations where the macroscopic state of a system changes qualitatively.

Let us consider a few examples. In physics an example is provided by the light source laser. When the laser is excited only weakly, it emits the light of a typical lamp, i.e. light consisting of many uncorrelated wave trains. If the energy input into the laser is increased, this microscopically chaotic light is suddenly replaced by a highly ordered light wave. When a fluid is heated from below, beyond a critical temperature difference between the lower and upper surface, patterns may be formed, e.g. in form of hexagons or rolls. Quadrupeds may change their gaits when a higher speed is required. Within synergetics, three different approaches have been developed to cope with these qualitative changes:

1. The microscopic approach. Here we start from the variables describing the individual parts of a system. The variables are lumped together into a state vector q which obeys nonlinear evolution equations. When control parameters are changed, a specific state q may become unstable and is replaced by a new state describing spatial, temporal, or functional structures. According to the slaving principle of synergetics close to an instability point, the behavior of the individual parts is determined by the order parameters. Instead of the need to describe the behavior of the individual systems, we need to focus our attention only on the behavior of the few order parameters. In this way, an enormous information compression is achieved. The behavior of the system is described by the order parameter equations, which in a number of cases can be put into classes. This allows us in particular to draw analogies between the behavior of systems that originally belong to quite different systems or even disciplines. The microscopic approach has proved useful in a number of physical systems, where one may start from basic equations for the individual parts, but also with respect to model equations introduced in chemistry, biology, and other fields. These approaches have been described elsewhere in great detail [1], [2].

2. The approach of macroscopic synergetics [3]. In many cases, the parameters or variables describing the individual parts of a system are not well-known or not known at all. On the other hand, measurements on some macroscopic properties of the system can be performed, for instance one may measure some macroscopic variables and their moments up to a certain order, as well as some specific correlation functions. In such a case, unbiased guesses on the underlying dynamics can be made by use of the maximum information (entropy) principle or generalizations of it. The details of this procedure have been worked out in

the case that the underlying dynamics is Markovian [4]. Explicit examples so far treated refer to a Brownian motion in a nonlinear potential (up to cubic order) and some two-dimensional cases. It is expected that this procedure will allow us to analyse time series from a new point of view.

3. The phenomenological approach (cf. [5]). This approach starts from the fact that close to instability points the behavior of a system is governed by few order parameters. Instead of deriving the order parameter equations from microscopic equations one may try to write down such equations directly. An example is provided by the modelling of the finger movement experiments done by Kelso [6]. Kelso's experiment, which has become a paradigm for a whole class of related experiments, is as follows: Test persons are asked to move their fingers in parallel at an increasing speed. Beyond a certain speed, the parallel finger movement is changed involuntarily into a antiparallel, i.e. symmetric, movement. All the phenomena known from nonequilibrium phase transitions [1], namely critical slowing down, critical fluctuations, as well as the qualitative change of the behavior could be treated by a simple model for the order parameter represented by the relative phase of the two fingers [7], [8]. This experiment and its model is particularly remarkable, because quite obviously any human is an extremely complex system. Nevertheless, it has become possible to model specific kinds of changes of behavioral patterns in great detail by the order parameter concept. Another example is the analysis of EEG and MEG patterns. In such a case, the analysis of multielectrode derivations in the α-wave region showed that the underlying spatio-temporal potential fields can be conceived as superpositions of few basic modes only. In the case of epileptic seizures, these modes could be identified including their dynamics described by three order parameters whose equations could be written down explicitly [9], [10]. In the case of normal behavior of people in a resting state with their eyes closed, the spatio-temporal pattern can be represented as a superposition of five basic spatial modes. Their amplitudes are again the order parameters. In this case, however, the underlying dynamics could not yet be modelled by means of simple order parameter equations. It is hoped, however, that by the inclusion of the methods quoted above under 2., such an analysis may become possible.

3 Order Parameters and Parts. Some Examples from Sociology

The typical relationship between order parameters and the individual parts of a system can be found in many disciplines including the humanities. Let me give a few examples: Language is quite evidently an order parameter. It lives much longer than any individual of a nation. Once a baby is born, he or she is subjected to the language of his or her parents. The baby learns the language and then carries the language further. We here observe the same relationship between the order parameters and the individual parts as characterized by circular causality. Another example is provided by rituals and the individuals obeying rituals. In our context, the meaning of a ritual is the marking of a consensus between people or the marking of the identity of a group. Other examples for order parameters are nations that are supported by their citizens, or law that, at least in a democratic country, is produced by its citizens. Another example for an order parameter is described by the climate in a company or by corporate identity. The climate of a company is formed by the individual members of a company who, in turn, are determined in their behavior by that climate. Another and less serious example for an order parameter is fashion and in what way it determines the kind of dresses people wear.

There are far more serious examples for order parameters, however. For instance, scientific theories may be considered as order parameters, or as Thomas S. Kuhn called them, as paradigms. I refer to his book "The Nature of Scientific Revolutions". Scientific theories are carried on by scientists and then the students are taught these theories. In that way they are, if one wishes to say, "enslaved" by theories or concepts. Another order parameter is the kind of economic system that determines the behavior of producers and consumers and in turn is determined by the behavior of the latter. It is worthwhile mentioning that prominent economic scientists, such as Samuelson, already distinguished between slow and fast variables in economic processes. But, of course, they had no mathematical tools to formalize these concepts during their time.

In all these examples the order parameters seem to be rather rigid and one may wonder how one may change order parameters, e.g. the climate in a company, or an economic or political system. And what phenomena are accompanying that changes? Synergetics has elucidated the mechanisms by which such changes may occur, and I refer here again to the laser paradigm. In all cases, where self-organization occurs, or is wanted to occur, we cannot directly determine the behavior of the individual parts. Rather we have to change unspecific control parameters. Let me demonstrate these ideas by a rather extreme example which can be nicely illustrated by means of the hilly landscape I have used before [1]. How can one go from one economic system to another one? First the landscape must be deformed, i.e. one has to destabilize the old system. This may be achieved, for instance, by loosening strict regulations, by allowing other kinds of monetary fluxes than before, a.s.o.. Then, eventually, a new system may evolve as characterized by newly appearing minima. But there are several important features to be noted: one runs through a period of critical fluctuations and critical slowing down, a phenomenon which is well observed in some economies at present, e.g. the former Soviet Union, and later on we have to expect the problem of symmetry breaking. I.e. according to the laws of synergetics, we must not expect that a destabilized system will automatically run into a specific new state. Quite often there are several possible new stable states, some being optimal, some being suboptimal, and small fluctuations, for instance the action of a small group of people, determine the course of the events.

4 Stability versus Adaptability

Within the microscopic approach of synergetics it has been established that close to instability points, where the macroscopic behavior of a system changes qualitatively, the dynamics is governed by few variables, the order parameters. Taking this result for granted, a number of complex systems become accessible to be modelled close to such critical points from where one may extrapolate into noncritical regions. In the case of sociology, this approach has been proposed by Wunderlin (compare these proceedings), because the complex behavior of individuals can hardly be modelled adequately. On the other hand, this kind of modelling is not necessary once we deal with an entity of people in situations where qualitative changes at a macroscopic level of a society occur. This approach of synergetics is also interesting from a more philosophical point of view. While in earlier times science was mainly interested in stable states, it is more and more focussing its attention to unstability points. Also from a practical point of view, it is important to stress that in order to keep systems adaptable, they should be kept close to instability points from where they can more easily and rapidly adjust to a new situation. The comprehensive physiological studies by Koepchen [11] show clearly that considerable fluctuations of physiological data, such as blood pressure, heart beat, etc. occur, in particular when persons are in a resting state. It is as if the person is testing virtually possible new states to be able to adapt quickly to new situations. Such a concept of adaptability by means of critical points has far-reaching consequences for instance for the treatment of an economy or of social systems. In order to be adaptable, the system must not be too stable, rather it should allow for test processes in form of fluctuations close to critical points.

The visualization of the behavior of a complex system by means of the movement of a ball in a hilly landscape may serve us also to discuss the relationship between stability and adaptability. When there is only one valley with deep slopes, the ball is in a very stable position. Any small pushes won't move it around appreciably. However, there may be situations where softer slopes are more desirable. Consider a landscape with two valleys, one being deeper than the other one, and let us assume that the deeper minimum represents the state of a system with a higher efficiency. If the system is originally in the upper minimum, it cannot jump spontaneously into the lower one unless it is driven there by fluctuations. This may serve as a metaphor for many complex systems including those in economy. We must allow systems to adapt by means of fluctuations and the possibility that these fluctuations may even grow up. This picture seems to me to be at the heart of creativity. We must be able to let our mind diffuse around so that it, eventually, can conceive entirely new ideas. In a number of cases it may be desirable to change the behavior of systems between stability and adaptability. Nature shows us how adaptability is obtained by means of her evolution. Here the fluctuations are represented by mutations which in a favorable environment can give rise to new species. The same relationship between stability or even rigidity on the one hand and evolution on the other hand may be found in the pair ecology and technology. For me there is not the slightest doubt that mankind needs technology to be able to cope with the needs which are coming up time and again and to keep mankind adaptable enough to cope with all the difficult problems. Quite evidently, we have to find a balance between ecology and technology, or, in other words again, between stability and adaptability. There is certainly not a simple recipe how to obtain that balance, rather we have to strive to that goal. Let me pick up here one important aspect: In a way, nature is our great teacher from whom we can learn important things. One is, quite evidently, the aspect of recycling. When we look, for instance, at a forest, trees grow, die and then are recycled in the soil to give food for new trees. Matter is preserved in all these cases. What is needed in addition is the energy provided by the sunlight to the trees and the information in their genetic code. Thus, quite evidently, we have to strive after the exploration of new energy sources and the use of more and more information or knowledge. In his studies, Ake Andersson in Sweden has quite clearly identified knowledge as one of the most important order parameters in economy. This is, quite obviously, a trend which has to be followed up. We have to replace material by information. Since our energy resources are limited, we have also to replace energy by information.

Minimizing the exploitation of energy sources means that we have to increase the efficiency of processes. Again the laser provides us with a beautiful example: When it acquires its coherent operation, its efficiency rises dramatically. At a more abstract level, we may say that the ordering of the laser electrons is brought about by a sufficiently strong exchange of information. But there are profound differences between a laser atom and a human being. The difference lies not only in the complexity of humans but also in the ability to learn and to transfer their knowledge to further generations.

According to synergetics we may change a state of a self-organizing system by indirect means. These means may be quite subtle; even small changes of "control parameters" may induce dramatic changes of the whole system. They may lead to an improvement of the system, but even a slight change of a control parameter in the wrong direction may lead to serious difficulties. We are becoming aware of how fragile complex systems are.

References

1. H. Haken, Synergetics, An Introduction, 3rd. ed., Springer, Berlin (1983)
2. H. Haken, Advanced Synergetics, 2nd. print., Springer, Berlin (1987)
3. H. Haken, Information and Self-Organization, Springer, Berlin (1988)
4. L. Borland and H. Haken, Learning the Dynamics of Two-Dimensional Stochastic Markov Processes, to be published (1992)
5. W. Wischert, A. Wunderlin, On the Application of Synergetics to Social Systems, these proceedings (1992)
6. J.A.S. Kelso, Phase transitions and critical behavior in human bimanual coordination, Am. J. of Physiolog. 246, R1000-R11004 (1984)
7. H. Haken, J.A.S. Kelso, H. Bunz, Biol. Cybern. 51, 347-356 (1985)
8. G. Schoener, H. Haken, J.A.S. Kelso, Biol. Cybernetics, 53, 442 (1986)
9. A. Fuchs, R. Friedrich, H. Haken, D. Lehmann, in: Computational Systems - Natural and Artificial, H. Haken (ed.), Springer, Berlin (1987)
10. R. Friedrich, A. Fuchs, H. Haken, in: Synergetics of Physiological Rhythms, H. Haken and H.P. Koepchen (eds.), Springer, Berlin (1991)
11. H.-P. Koepchen, these proceedings with further references

Evolution, Creativity and Intelligence in Complex Systems

*P. M. Allen and H. K. Phang**

International Ecotechnology Research Centre,
Cranfield Institute of Technology,
Bedford, MK43 OAL, England

1. Introduction

In order to make "intelligent" decisions we need to act "successfully", and the only indisputable measure of "success" is of course that of survival in the arena in question. Action is taken on the basis of two beliefs. First, we must either believe that we know the possible outcomes of our actions (or know to what extent we cannot know what those outcomes will be), and second, we must have formulated some values or goals by which we believe we can measure the expected "performance" of our choice.

Unfortunately, until now the conceptual and mathematical basis for such wisdom has not existed. In mathematical models of human systems, the values and goals of people were assumed to be "given", and the models have focused on the issue of prediction. Predictive mathematical models are then based either on differential equations, representing the actions of fixed mechanisms, excluding learning and adaptation, or worse still are based on assumptions of equilibria and optimality. These methods are really ideas that were based on the false analogy between an *isolated* physical system which can be shown to approach an equilibrium state, and the complex evolution of a socio-economic system over time.

The traditional scientific view has been that of the mechanical paradigm in which a system is assumed to have a trajectory along which it moves "naturally" as a result of the continuous action of the mechanisms which operate between its components. In addition, with frictional forces, such a system is assumed to move from its initial condition towards a new "equilibrium" state, in which dissipation has either ceased or is minimized. The cause of change is therefore viewed as being outside the system under study, and the final state is predictable as that of thermodynamic equilibrium.

But although this view is adequate for closed, physical systems, it is totally inadequate for complex systems such as those containing living beings, where evolution and change run mostly toward increasing complexity and organization, and in which creativity and learning play a vital role. For these, a new approach is required in which the system evolves as a result of a continuous co-evolutionary dialogue between the microscopic and macroscopic levels of description of the system. Indeed, creativity itself seems to result from the necessarily imperfect understanding of the

* *Researcher, Toppan Moore Systems Limited (Japan)*

Springer Series in Synergetics, Vol. 62 **Interdisciplinary Approaches to Nonlinear Complex Systems** – Eds.: H. Haken and A. Mikhailov

system that each individual element must have, so that goals, strategies and perceived pay-offs are sufficiently unclear to allow exploratory behaviour, and in this way link together the evolution of the "character" and "beliefs" of individuals with the larger structures which they fashion.

2. Complex Systems - Multi-Dimensional Evolution

In the past, the social sciences have preferred to try to represent human decision making in terms of "rational" and perfect, or possibly slightly imperfect information. However, the view which we shall develop in this paper, is that of very "bounded" and "imperfect" rationality, but we shall argue that this is actually creative and part of evolution.

In economics, for example, people are assumed to know not only the present situation (which they may possibly have some idea of), but also the results of all possible experiences that have or have not taken place, so that they are in a position to choose from all these possibilities the action that best fits their particular goals. For example, the "supply" and "demand" curve, probably the most basic diagram in economics, assumes that at any given time there *is* a curve which expresses for any price how much of a good would be produced or bought. Yet obviously, this is in reality very doubtful, since it depends on the strategies of different producers, responding to their competitors actions, and for consumers it depends on fashion, availability and service, which depends itself on the quantities sold.

The fact is that individuals do *not* necessarily *understand* the world around them. Why should they? How could they? Survival and evolution only require that no fatal acts have been committed (yet), but in no way implies that survivors should have a perfect vision of the world and its workings, or indeed that *such a vision exists*. Collective systems have multiple aspects: psychological, informational, social, material, cultural, historical, epidemiological, economic, etc. And, despite the fact that all of these different areas of interest are merely facets of the complex system that has arisen, the human mind has no choice but to separate them out, into a "taxonomy" of disciplines for further study. But that alone is not the problem, it is that they are usually studied *separately*. So, for example, economics, social science, psychology etc. are considered as academic "subjects", worthy of isolated study. What may have begun as a simple desire to specialize in a particular area, within the context of the others, has led instead to a disconnection and separation of "expertise" to the point where this is often the main problem. Figure 1 illustrates the complex system of economics.

If we accept the separate study of subjects like "economics" for example, then it means that we must believe that the flows of goods and factors, and the prices that are observed can form some kind of closed self-consistent system in themselves and explanation can be found from within this narrow structure. Since, in reality, strong interactions traverse the boundaries of economics from the psychological, technological and environmental domains to name but a few, the only way that "explanation" could be found "within" economics is if there were some set of fixed relations between the variables which held over time. But the evidence of experience is that these do not exist in reality, and the emergence and development of economic systems, is a story of real, structural changes, where innovations, changing beliefs and values characterise

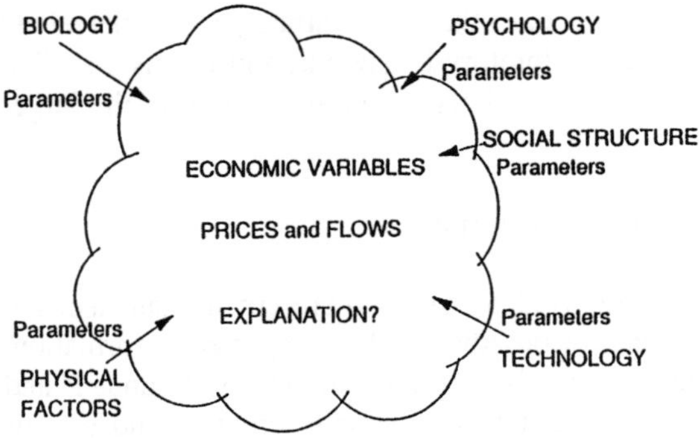

Figure 1. Economics is part of a complex system. Explanation within the system is false.

the system, and restructure it. We must look to a new multi-disciplinary, evolutionary complex systems approach for a new understanding.

These ideas have already been presented in several papers by one of us (Allen, 1988, 1990, 1992) and will not be repeated here. In essence, the evolution of complex systems is viewed as occurring on two levels of description. There is the macroscopic, average level in terms of differential equations governing average behaviour, and the microscopic level of individual experience and necessarily incomplete knowledge, which underlies this. In systems with non-linear interactions however, the two descriptions can diverge, as fluctuations in microscopic behaviour are amplified and lead to an instability in the macroscopic structure of the system. As has been amply described in numerous articles and books, the evolution of non-linear systems proceeds through successive instabilities, as the space of possible "attractors" changes with some applied parameter. However, although such an approach is considerably richer than that of equilibrium, or of linear dynamics, it still relies on the myth of the "mechanical paradigm". This, for example, has been the approach of System Dynamics and now of many others interested in such phenomena as "deterministic chaos", in which the system is represented in terms of non-linear differential equations.

But this is inadequate in two respects. First it merely describes the "functioning" of a system with fixed mechanisms, and not how it became what it is, nor how it might become what it is not. Secondly, because it assumes fixed average mechanisms of interaction between individuals, it excludes the possibility that creativity, learning and intelligence on their part might affect the system as a whole.

In this paper we wish to present an approach which overcomes these two inadequacies. We shall represent the mutual coevolution of different individual "strategies" with the macroscopic, average structures which they both create and inhabit.

3. Evolutionary Drive

In order to address the problem of the mutual evolution of the nature and beliefs of individuals within a system, and the collective structure which they fashion, some

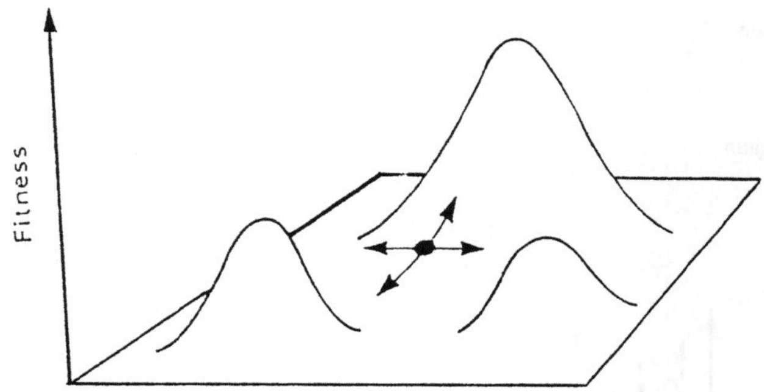

Figure 2. In "possibility space", the behaviour of a group will spread outwards as a result of imperfect reproduction and learning.

complex simulations have been developed (Allen & McGlade, 1987a,1989) examining the evolutionary dialogue between "average" (macro) processes and the "non-average" (micro) detail. This led to the new concept of *evolutionary drive.*

The interesting idea here was to define a "possibility space", as shown in figure 2, a space representing the characteristics, behaviours and beliefs that could potentially arise for the different types of individuals present. In practice, of course, this is a multi-dimensional space of which we would only be able to anticipate a few of the principle dimensions. However, it is nevertheless extremely instructive to think about the evolutionary process in these terms. The central problem of change is that of understanding how, over time, the kinds of behaviour present in a system can actually increase and complexify. In terms of "possibility space", we may say that if initially there is a single type of individual present, occupying a single cell of this space, then how can new populations appear?

The answer clearly, is that this "possibility space" will be explored by individuals if their behaviour is plastic in some way. In biological evolution, we know that not only are there mutations, but more importantly sexual reproduction leads to the production of off-spring which are not exact copies of either of their parents. The genetic mechanism is precisely such that a large range of possibilities are explored, and off-spring, off-spring of off-spring and so on, spread out over time from any pure condition. In human societies, possibility space is explored by the variety of encounters and meetings that can occur, imperfect information and learning, together with individuality. The differential success caused will reduce the spread, and make average behaviour improve.

Physical constraints automatically ensure that some behaviours do better than others, and so there is a differential rate of survival and of production. If possibility space is seen as a kind of "evolutionary landscape", with hills representing behaviours of high performance, then our simulations lead to the amplification of populations which are higher on the hill, and the suppression of those which are lower down.

The initial papers showed how the *imperfect reproduction* and error making of populations provided a capacity to climb the hills of a fixed adaptive landscape. A

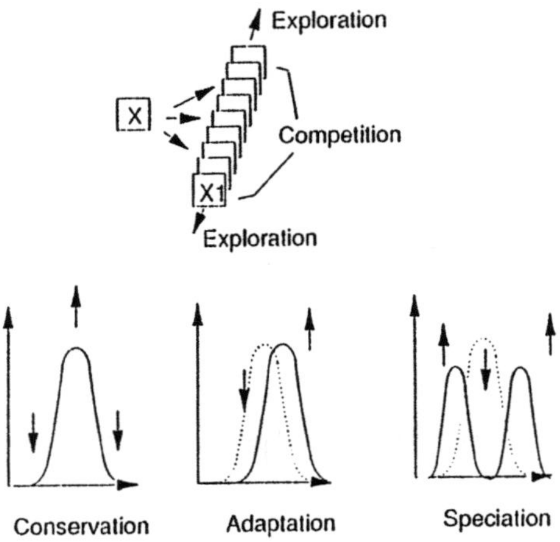

Figure 3. Evolution results from the spreading of a population distribution in character space, and the different success of each type.

display of "intelligent" adaptation was shown to require diverse, error-making behaviour in a population, and therefore to be necessarily sub-optimal at any given time during the process. Furthermore, by examining the competition between populations with different intensities of "error-making" or "exploration" it was shown that the steeper the slope the more advantageous was error-making. In a fixed landscape, the inevitable conclusion would be that once a population had climbed to the summit of a hill, then error-making would be of negative value and would be switched off, together with the capacity for an "intelligent" response.

But,in reality,the hills of the adaptive landscape are not fixed, they are created by the behaviours of the other populations present in the system, and because of this, error-making and the accompanying capacity for intelligent response can never be turned off, since the co-evolution of the different populations will never come to an end. Evolution, instead of leading to the creation of individuals with "optimal" behaviour, leads to populations with the capacity to learn, and in consequence diversity and sub-optimal behaviours on the part of individuals with an imperfect understanding of the system that they inhabit.

Successful evolution is a seemingly suboptimal, messy process resulting from the interplay of *exploratory diffusion* of individuals in some behaviour space, and the selection operated by their *differential success.* There is a process of simultaneous "stretching" and "squeezing" of populations in the space of possible behaviours, as shown in figure 3, that is at the core of our new understanding.

Simulations of this "evolutionary drive" have shown explicitly that where behaviours happen to provoke positive feedback loops of interaction which favour themselves,then these are the ones which will grow in the system. Because of this, evolution is not simply a smooth process of continuous adjustment, but instead will move in steps as positive feedback locks in particular features, and resists structural change for some time. This result agrees with the hypercyclic models of Eigen and Schuster, which they use to describe the origin of life.

The equations which have been used in our simulations are essentially logistic growth equations, with error-making in the two dimensional possibility space that we have imagined. Different locations within this space represent different resource-acquiring behaviours, skills, and as a first approximation :-

$$\frac{dx_{(i,j)}}{dt} = bx_{(i,j)}\left(1 - \frac{x_{(i,j)}}{N}\right) - mx_{(i,j)} \tag{1}$$

Where :- $x_{(i,j)}$ = population with a particular behaviour
b = the rate of population increase
m = the rate of population decrease
N = the resources available to $x_{(i,j)}$

Diffusion within this space represents mutation of behaviour, f = fidelity of reproduction. Different behaviours are centred on different parts of the resource spectrum. Populations constrain each other's growth as a function of their size and separation owing to overlapping consumption of resources, ρ = decay factor in resource overlap with distance, d = distance from $X_{(i,j)}$ to $X_{(i',j')}$.

Provision is made for the tendency of certain characteristics to increase the probability of their occurrence, w = rate of positive feedback. With these added details the equation becomes :-

$$\frac{dx_{(i,j)}}{dt} = [(fbx_{(i,j)} + wx^2_{(i,j)}) + \frac{b(1-f)(x_{(i+1,j)} + wx^2_{(i,j)})}{4} + \frac{b(1-f)(x_{(i-1,j)} + wx^2_{(i-1,j)})}{4} + \frac{b(1-f)(x_{(i,j+1)} + wx^2_{(i,j+1)})}{4} + \frac{b(1-f)(x_{(i,j-1)} + wx^2_{(i,j-1)})}{4}] \times [1 - \frac{\sum_{(i',j')} x_{(i',j')} e^{-\rho d(ij\,,\,i'j')}}{N \sum_{(i',j')} e^{-\rho d(ij\,,\,i'j')}}] - mx_{(i,j)} \tag{2}$$

A typical evolution that the model gives rise to is shown in figure 4, where an evolutionary tree emerges endogenously from the model. Appendix A shows a two-dimensional view of the gradual emergence of different populations over time.

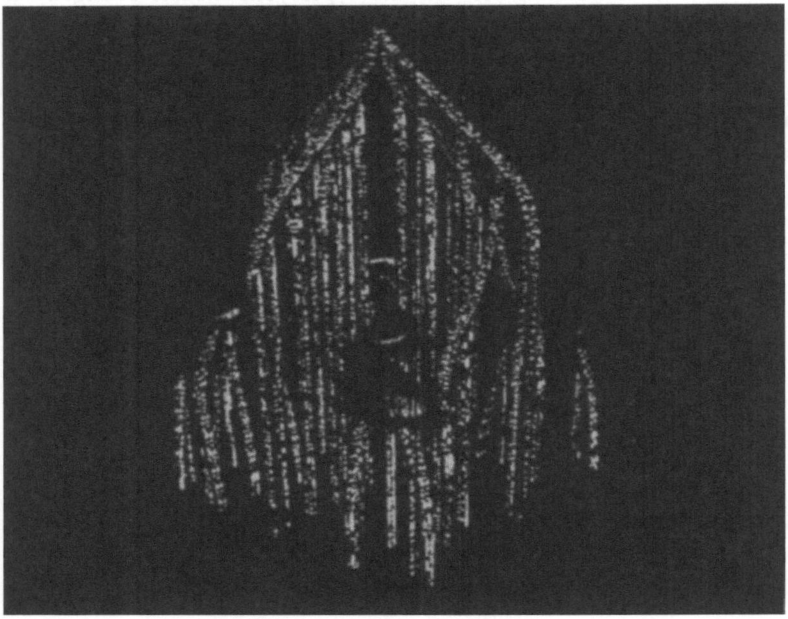

Figure 4. A 3-D visualization of our model of emergent structure. An ecology of interdependent behaviour emerges.

From a single activity, our model generates a simple ecology, and a dynamic one since the "identity" of each behaviour is maintained by the balance between a continual diffusion of deviants outwards into character space, and the competitive field that exists around it. Random events which occur during the "filling" process will affect *which* populations arise, and so it is not true that the evolution represents the discovery of pre-existing "niches".

Such a system operates *beyond* the mechanical paradigm, because its response to external interventions can involve changes in structure and of the "nature" of the activities or technologies in the system. Suppressing particular activities in such a system, as a result of some environmental change for example, will provoke a complex response from the system, as other activities adjust.

The particular pattern of clusters that emerges depends on the accidents of system history. The precise paths and direction of "explorations" are clearly very random, and so the pattern may be essentially unpredictable (Allen, 1990). Although the "inventiveness" of the population is constantly present, as there is diffusion into the possibility space, it is fascinating to see that our experiment shows that only at certain moments in time does this lead to structural change. In other words, the system evolves in phases of apparent stability, separated by periods of instability and fairly rapid re-organizations, although the disturbing pressure of exploration and creativity is relatively constant.

In human systems, this kind of evolutionary tree seems very common, and such positive feedback, hypercyclic, systems abound. Much of culture may well be behaviour which is fixed in this way, and we begin to see how difficult it really is to define "intelligence". In most situations imitative strategies cannot be eliminated by the evolutionary process, and so fashions, styles and indeed cultures rise and fall without necessarily expressing any clear functional advantages. Indeed, individual values and beliefs can be viewed not so much as leading to "the best" way of doing things somewhere, but perhaps as resulting from ignorance of other ways of doing things. Human activities in general exhibit these properties of autocatalytic, self-organization, where ritual and shared ideology emerge and serve as the identity and focus of a social group, irrespective of the precise merits or truth of the ideology itself. So much of human attention is focused on playing a role in groups where values are generated internally, and the physical world outside is largely irrelevant.

The ideas described above have already been applied in a number of fields:

a) the evolution of spatial distributions of economic activities and settlement patterns in the USA, Belgium, Holland, France and Senegal, as well as in understanding intra-urban evolution in cities of Belgium and France. (Allen & Sanglier, 1979,1981, Allen, 1985; Sanglier & Allen,1989, Allen et al 1992).
b) the management of renewable natural resources, including learning and technological change of fishermen and farmers (Allen & McGlade, 1986). (Allen & McGlade,1987b, 1987c)
c) the evolution of market systems.

This paper will focus on the case (c), and we shall explore the implications of the general ideas of "evolutionary drive" for our understanding of "intelligence" in the context of financial markets.

4. Intelligence and Learning in a Financial Market

Financial markets offer us an interesting, and archetypal example of a complex system. There is macroscopic level, for example for a variable like "price", and there is also a microscopic level of the many different actors attempting to anticipate changes in price, and searching for effective strategies to do this. But, despite this, the financial market is characterised by the "centrality" of the activity of buying and selling, and in this way avoids the greater complication of a spatial market system. It therefore provides a simpler, homogeneous environment for study whereby prices comprise the most important observable.

As mentioned above, the traditional paradigm of understanding in this arena has been built around the concept of equilibrium and the rational expectations theory. Also, because probability calculus was adopted as the mathematical basis of financial theories, it was not surprising that price distributions were assumed to be "normal". General equilibrium theory and efficient market hypothesis portray changes in the market as being due to external factors represented as a *random walk* (Osborne 1964,

Muth 1961, Fama 1970). The appropriateness of this approach has since been fiercely debated, partly because of the adoption of simplifying assumptions about the way investors behave as well as the correspondence of the resulting analytic framework to reality. Although statistical inference was able to provide a vast array of modelling and research tools, the limitations imposed by the underlying assumptions clearly under-estimated crucial and complex mechanisms that endogenously drive events within the financial system.

Many discrepancies or exceptional behaviours that contradict this approach have long been discovered and documented but only to be conveniently classified as "anomalies". Some examples are seasonal effect (Watchel 1942, Rozeff & Kinney 1976, Debondt & Thaler 1987), including the turn-of-the-year and day-of-week effect, as well as size effect (Blume & Stanbaugh 1983). Behaviours that cannot be accounted for in the equilibrium models are conveniently studied as exceptions to the normal and treated as irregularities. This convenient classification then dismisses completely that these *"anomalies"* can potentially turn out to be the consequences of *"normal"* processes.

The subsequent realisation that equilibrium forms only a partial view and indications of the existence of *instability* through market crashes has motivated the search for alternative paradigms. Nonlinear dynamics, in particular *chaos theory*, has inspired many in their research (Blank, 1991; Chen, 1988; Day, 1983; Grandmont & Malgrange, 1986; Hsieh, 1991, Kesley, 1988; Larrain, 1991; Peters, 1991a; 1989; 1991b; Savit, 1988; 1989; Scheinkman & LeBaron, 1989). In order to apply the mathematics, one needs to identify, as a first step, the existence of nonlinearities as well as chaotic behaviours in the time series concerned. Interestingly, most analytical results indicated that nonlinearities and chaotic behaviours do indeed exist marginally in most cases.

Although deterministic chaos itself is a very interesting phenomenon and has given rise to much exciting research, the implications of the earlier sections of this paper are that differential equations are in any case an inadequate description of the evolution of such systems, and that the underlying microdiversity of actors needs to be taken into account explicitly. In reality, the strategies used by different actors are also "produced" within the system and local variations are constantly testing the effects of possible modifications. Here we shall examine the evolution of successful strategies in "possibility space" of trading rules.

Rational analysis is the traditional scientific approach to problems, and it can deal successfully with systems which are knowable, providing that we know what it is that we would like to minimise or maximise, and of course we would only be able to know what would be good "objectives" if we knew and understood the system. But in contrast, evolution shows us how to deal with a system that cannot be known completely, and for which, in consequence, we cannot formulate clear goals.

The stable functioning of a market can only result from the establishment of diverse, varied and changing strategies which are complementary to each other instead of all participants focusing on a single optimal strategy. Such systems represent a new domain of organisation beyond the "mechanical" where the strategies of traders are mutually interdependent, the system has many possible responses to perturbations and where survival is related to the *capacity to change*. An important feature of this capacity to change is that it involves sub-optimal behaviours, imperfect information, mistaken inferences and the power of creativity.

Let us consider financial markets from these points of view. At first it seems quite clear that our objective is to make profits over some particular time scale, and that therefore rational analysis of the "system" is what is required. But, of course, everyone wants to make profits and therefore to buy "cheap" and sell "dear". So, the real difference between the players in the market concerns what they believe about the future value of what they buy or sell. Indeed, for one player to buy what another sells implies that they almost certainly believe opposite things about the future movement of its price, or possibly that they are operating on different time scales. The important point however, is that there are a multitude of ways in which one can arrive at a view concerning whether a stock should be bought or sold and at what price.

Trading strategies are quite naturally the subject of constant discussion and debate within financial circles. Mostly, positions and views are held as a result of personal history and experience and beliefs about how the world works. But of course trading strategies concern the *future*, that is they are about actions which should be taken in order that the future will be influenced in a certain way. And this in turn implies that we are seeking a strategy which can give good results despite the fact that we *cannot* know "the" future, because there are in fact different possible futures. When some trends in the markets become apparent, the traders will then react to this, and by their actions change what subsequently will occur in reality. This implies that markets will always drive themselves to the *"edge" of predictability* and therefore we should try to understand and learn to manage the processes of change rather than predict future prices.

Thus, it becomes quite clear that in order to deal with the "unknown" future successfully, or at least sustainably, diversity and adaptability in the strategies adopted are most important, rather than finding the single "optimal" strategy. Systems that simultaneously modify both the macro-structure as well as the nature of the underlying micro-components require more than the usual "mechanical" solutions that had been so excessively prescribed in the past. *Intelligence* in managing such systems, financial markets in this case, depends very much on the capacity to change and the time-scale of this adaptability is vital in determining success and sustainability. Similarly, the role of diversity should be emphasised for fear of falling into the positive feedback trap, in the event where success overshadow linkages to the external environment and subsequently, overlook the need to persist in the exploration of the continuously changing strategy space.

5. A Self-Adaptive Trading Model

In this particular work, a trading model was built to study the relevant issues raised in the last section. A popular and relatively elementary mechanism, the *moving average* method, was chosen to produce the appropriate trading signals. Despite the proliferation of analytical methods used in practice, the moving average is one of the most simple, yet versatile and widely used of all technical indicators (Murphy, 1986).

Because of the way it is constructed and the fact that it can be easily quantified and tested, it is the basis for many mechanical trend-following systems in use today.

A "moving average" calculates the average of the most recent values of an on-going time series. It has to be decided over how far back (usually in days) the moving average will extend. The body of data to be averaged moves forward with each new trading period. The period to be used is a nominal number that is subjective to each individual user and it is this number that, more often than not, determines the success of a strategy that employs the method.

Essentially, moving averages are calculated everyday and since these are figures based on the last n periods before, the line formed by connecting these averages will lag the actual price line which will cross it whenever a trend reverses. It is this *"crossing"* phenomenon that is taken to indicate a change in price trend, and hence a moment to buy or sell the commodity in question. The moving average is a follower, not a leader. It never anticipates, it only reacts. The manner of interpretation really depends on each individual and it is both the selection of the period n as well as the timing of each transaction that commands the success of such a strategy.

If the price of a commodity traces out a fluctuating path, moving unpredictably through peaks and troughs, then providing that the up and down trends can be detected early enough, profit can be made by applying the moving average method. It usually signals a change in trend when the price curve itself cuts the moving average curve. However, this method can also go astray if the price produces a "spike", cutting the moving average curve and then immediately cutting it in the opposite direction. Such an event will result in losses - not just by buying high and selling low, but also for trading costs.

A *bandwidth* can be added to this method to act as a precaution against unnecessary trades. This is a measure above as well as below the calculated moving average curve. Instead of transacting trades when the moving average curve cuts the original price curve, the bandwidth ensures that the moving average curve has to cut through it as well before it is interpreted as a signal to buy or sell. This reduces the number of occurrences whereby "spikes" would have initiated a false trade.

This means that the analogy to our "possibility space" of the previous section is some space of possible strategies. Figure 5 illustrates the two-dimensional strategyspaces of the NICKEL futures contract taken at different time intervals based on the moving average period and bandwidth parameter. The variability in the landscape caused by *payoffs* changing at different parts of the curve captures succinctly the problem that a static strategy will face if used throughout the period. The unevenness shows that a particular strategy has good and bad patches and unless adaptability is introduced, the strategy will not be sustainable.

Basically, the self-adaptive concept entails the division of the historical period into several equal sections, which we will call x. To start the simulation, the strategy first selects for the "best" moving average period n from a potential array of parallel strategies over section x based on the criteria set. This period n is then used in the trading simulation for the next period y. At the end of period y, the strategy looks back at that point in time into a period of x duration before and performs the selection process again. The trading simulation is then repeated for the next period y again and

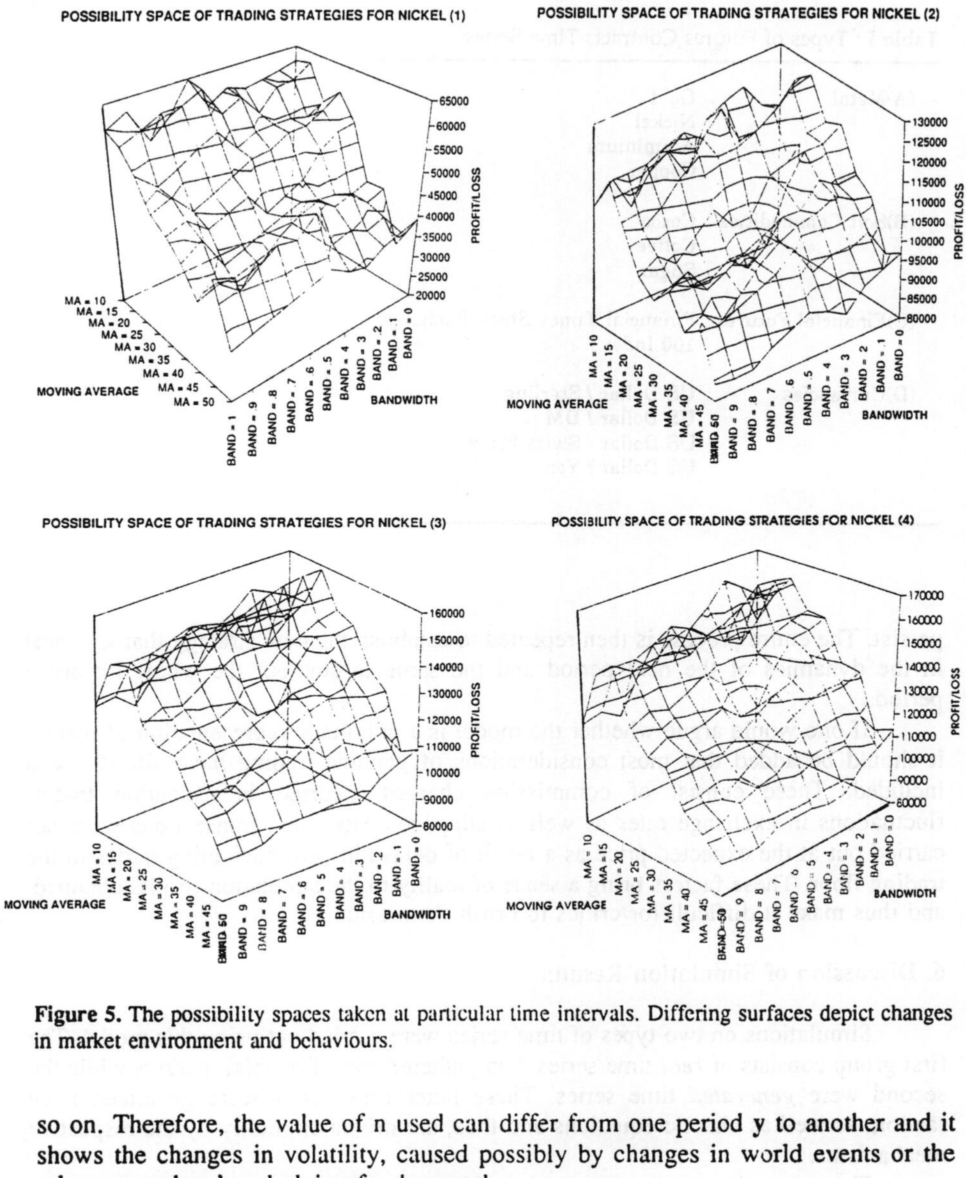

Figure 5. The possibility spaces taken at particular time intervals. Differing surfaces depict changes in market environment and behaviours.

so on. Therefore, the value of n used can differ from one period y to another and it shows the changes in volatility, caused possibly by changes in world events or the relevant markets' underlying fundamentals.

Besides selecting for the "best" strategy from an array of *parallel strategies*, the model also investigates for the appropriate time scales in which the strategy can most ideally operate in. x is really the length of time period in which one believes that the *current prevailing dynamics* presides. On the other hand, parameter y denotes the *persistence time* of the collective dynamics found in the period x before. In layman terms, a triumphant strategy found in a period x will be used in the coming y period as that is the time horizon that the prevailing dynamics of the market is found to

Table I : Types of Futures Contracts Time Series

(A)Metal	Gold Nickel Aluminium Copper
(B)Soft Commodities	Cocoa Coffee Sugar
(C)Financial Futures	Financial Times Stock Exchange 100 Index
(D)Currencies	US Dollar / Sterling US Dollar / DM US Dollar / Swiss Franc US Dollar / Yen

persist. The entire process is then repeated to establish the new strategy that will deal in the dynamics of the next period and the same throughout the whole historical period.

If one would argue whether the model is a pragmatic representation of reality, it should be added that most considerations of genuine trading have already been included. These consist of commission charges incurred in executing trades, fluctuations in exchange rates as well as slippage costs when transactions were not carried out at the expected price as a result of delays in communicating trades to the trading floor. These factors bring a sense of reality to the simulation results obtained, and thus make it difficult for critics to brush them aside.

6. Discussion of Simulation Results

Simulations on two types of time series were carried out using this model. The first group consists of *real* time series data gathered from financial markets while the second were *generated* time series. These latter time series were generated from random as well as chaotic equations and thus, are driven explicitly by the respective mechanisms.

First, we will discuss the results of trading in the real market environment. These time series span across a spectrum of futures contracts, from metals, softs, financial index, to currencies, each over a total of 44 months expiring on the 31st December 1991 (see Table I). Out of the total of 12 futures contracts, 8 performed significantly better than an arbitrarily selected *static* strategy of 50-day moving average. The static strategy does not possess the adaptability property and as its name suggests, the same strategy is maintained in the entire period, including bad periods. The results also showed that without using any information from the future, which

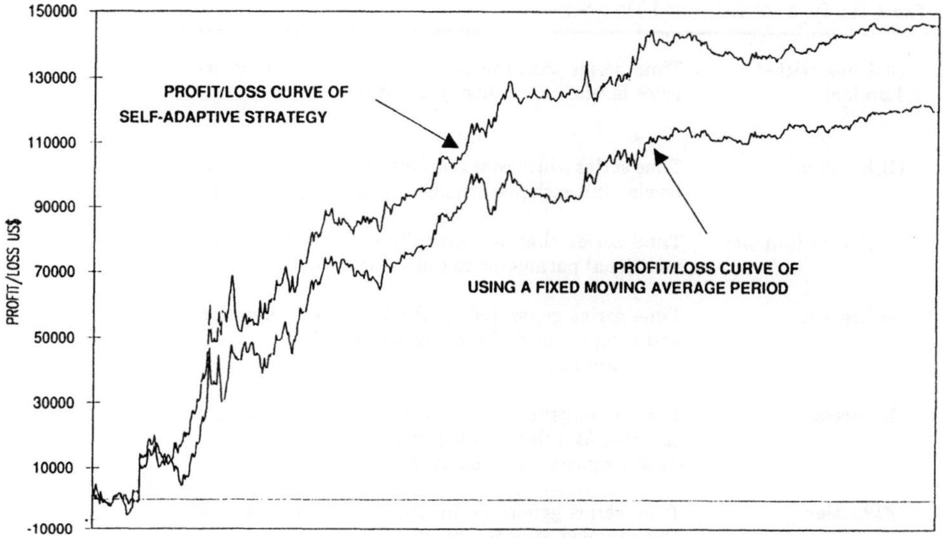

Figure 6. A comparison of performances obtained by the self-adaptive model and a static strategy in the NICKEL futures contract.

many optimisation techniques do, it was still possible to obtain good performance (see figure 6).

The characteristics of the various time series had a significant impact on the use of strategies at different times which indicated that the model was indeed adjusting to changing conditions. This concept, with the strategy **co-evolving** with changing dynamics in the markets, clearly corresponds to the basis of the evolutionary framework

Moving on to the group of generated time series, 6 varieties were generated and tested. Table II is a list of the various types of time series used. The simulation results displayed the similar improvement over the static strategy as the first group of real time series. Only a minority of 4 time series produced worse results in the group of 18. Unlike the real time series, where noise are present, these generated time series are purely driven by the underlying generating mechanisms used. Random walks of various kinds as well as chaotic equations were used to generate these price time series. This simulation exercise will reveal the evolutionary concept's capacity to deal with time series that are supposedly unpredictable.

By being able to deal successfully with a unknown future, the trading model developed has shown its resilience towards unpredictability. The potential diversity of strategies captured in the model lies in the form of 50 or so variations of moving average method. By reinforcing the successful strategy, the model adapts to changing price dynamics and in the process produce better performance. Thus, the model shows us that the evolutionary approach has the capacity to deal successfully and sustainably with systems that are faced with an unknown future. From this, intelligence is not the

Table II : Types of Generated Time Series

(A)Unrestricted Random	Time series that contains no memory of previous price levels, i.e. randomly generated.
(B)Random	Time series which contains memory of previous price levels but each price change is generated randomly.
(C)Biased Random	Time series that is randomly generated with an additional parameter to control its trend.
(D)Logistic	Time series generated by the LOGISTIC equation and sampled at different time intervals. $x_{(t+1)} = ax_t(1-x_t)$
(E)Lorenz	Time series generated by the LORENZ equation and sampled at different time intervals. dx/dt=sigma(y-x); dy/dt=rx-y-xz; dz/dt=xy-bz
(F)Rossler	Time series generated by the ROSSLER equation and sampled at different time intervals. dx/dt=-(y+z); dy/dt=x+ay; dz/dt=b+z(x-c)

ability to achieve optimal performance in the short term but the capacity to know how and when to change its strategy. In systems where events are dependent on the strategies of surrounding populations, continual modifications of views and a process of imperfect learning compose the only form of intelligence.

As an interesting aside, the model can be extended to further study the effects of positive feedback by channelling the impact of trading activities directly onto price returns and the subsequent altered course of events.

7. General Discussion

In this and earlier papers, it has been shown that the evolution of complex systems is a process that goes beyond the "mechanical" paradigm. Instead of the system being viewed as a "point" moving on a trajectory in some fixed landscape of attractors, we see that not only is the landscape itself generated by the actors in interaction, but also that the system itself can never be reduced to a "point" in phase space. The macroscopic description in terms of differential equations is an approximation to reality, within which there is an underlying microscopic diversity, which explores the stability of the "taxonomy" that has led to the choice of variables in the model. More importantly, not only is this micro-diversity important, but it is also maintained and fed by "error-making" and imperfect information at the individual, microscopic level, which confers creativity and seeming intelligence on the system as a whole.

Because of creativity, and learning, the only *certainty* in an evolutionary system is *uncertainty*. At any particular instant imperfect information concerning "what to do" is inevitable if the system is creative, and naturally, creativity is in turn inevitable if the individual, micro-level components do not know exactly what they should do for the best, and indeed what factors define "best". Long term success is not just about improving performance with respect to the external environment of resources and technology, but also is affected by the "internal game" of a complex society. The "pay-off" of any action for an individual cannot be stated in absolute terms, because it depends on what other individuals are doing. Strategies are interdependent.

Ecological organization is what results from evolution, and it is composed of self-consistent "sets" of activities or strategies, both posing and solving the problems and opportunities of their mutual existence, as a result of varying views and initiatives which, when positive feedback outweighs negative, leads to a new feature in the system. Value is assigned afterwards as a "post-hoc explanation" rationalizing events by pretending that there was some pre-existing "niche" which was *revealed*, although in reality there may have been a million possible niches, and just one in particular arose.

The future, then, is not contained in the present, since the landscape is fashioned by the explorations of climbers, and adaptability will always be required. Intelligence is the "cognitive" response to this demand, and the problem we face is that of making successful, rather than unsuccessful adaptations over time.

How then can we learn in an uncertain world? The answer is through the "evolutionary approach". The weighting attached to different possible visions of the world and hence to possible strategies must be reinforced or diminished according to their relative success. Our financial model, demonstrates that even in chaotic systems, although short term "learning" is impossible, our trading model can still make a profit, since this is equivalent to learning in a statistical, longer term sense. Because the chaotic attractor is fixed, in our examples, the adaptive system can find parameters which trade successfully. If, on the other hand, the chaotic system were changing over time, then it would perhaps be more difficult to succeed, and in fact, the chaotic attractor would change if any significant actors succeeded in "learning" which parameters were successful. This is what was meant earlier by saying that the system runs itself on the "edge of predictability". The contest then is between actors with different "speeds" of learning, so that money can be made even in a chaotic system, provided that one continues to learn and adapt.

Evolution in human systems is a continual, imperfect learning process, arising out of the transfer from unsuccessful to more successful strategies, but never providing enough information for a complete understanding. Instead of the classical view of science eliminating uncertainty, as it reveals the working of the "machine", the new view today of complex systems accepts uncertainty and change as inevitable, and "evolution" as the response of the system to this reality. The evolutionary process therefore offers us the basis on which we may build "intelligent" systems that can help us better deal with our increasingly unpredictable and complex world.

Appendix A

The figures here show the successive emergence of populations over time as a result of running equation (2) starting from a single seed. The time interval between pictures is 400 generations, and we see how the taxanomy changes over time finishing with that of a simple ecology. The 2-dimensional space represents two dimensions of difference in behaviour or strategy which provide access to resources. The final picture shows the evolutionary tree of the whole process.

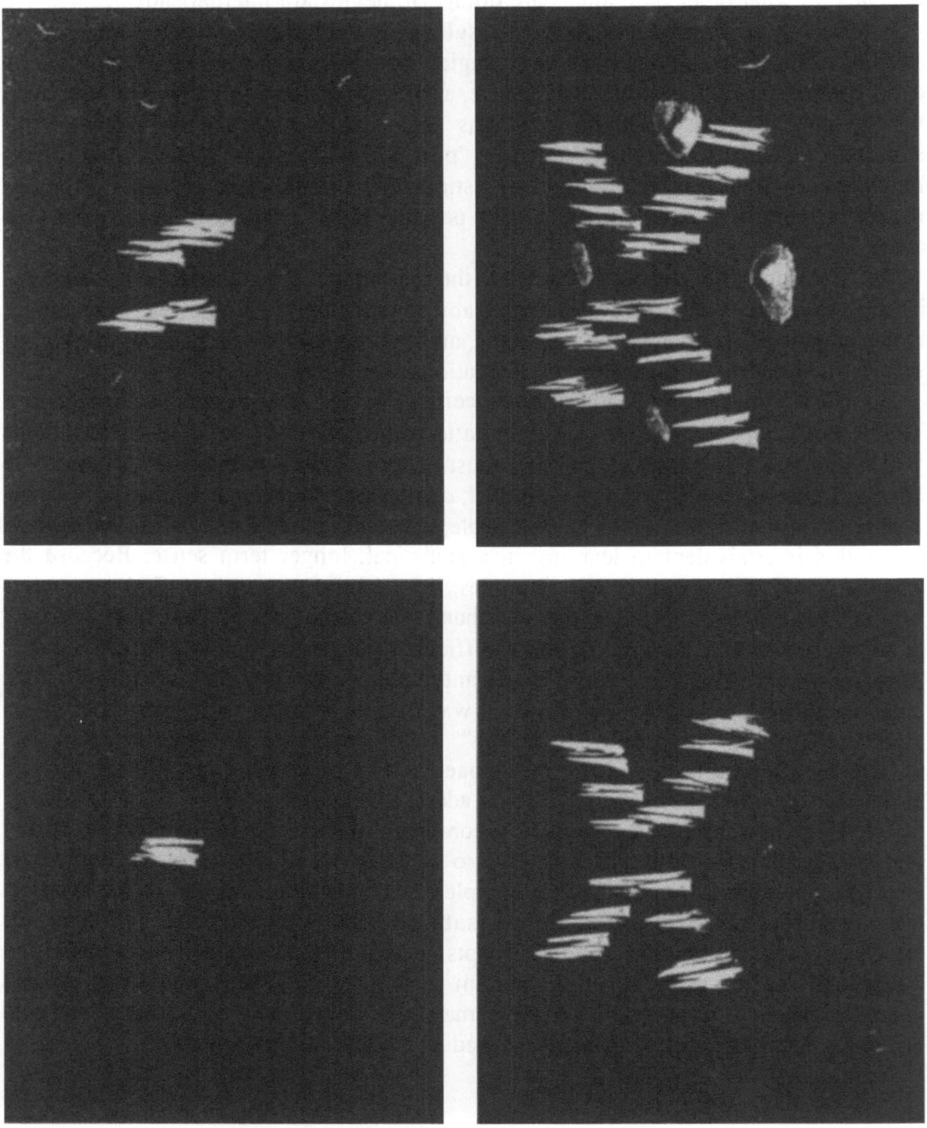

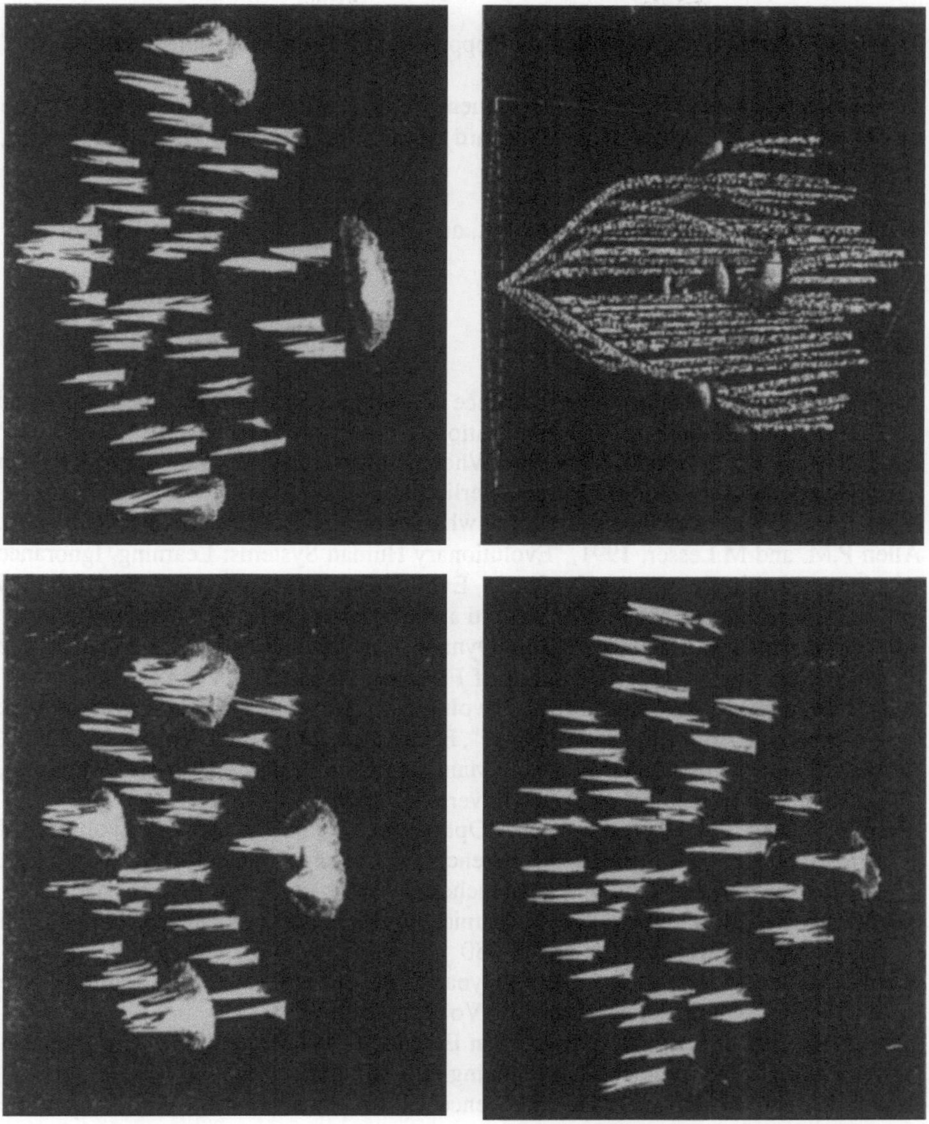

Acknowledgement

This work was partially supported by Toppan Moore Systems Limited (Japan).

The evolutionary tree (figure 4) and sequence in appendix A was produced by Mike Lesser and Dr. J. Corliss, at the Goddard Space Flight Centre, NASA, Greenbelt, Maryland.

We are grateful for the support of NASA, and also of the Open Society Fund of New York.

REFERENCES

Allen P.M., 1985, "Towards a new Science of Complex Systems, in "the Science and Praxis of Complexity", United Nations University Press, Tokyo.

Allen P.M., 1988 "Evolution: Why the Whole is greater than the sum of its parts", in Ecodynamics, Springer Verlag, Berlin.

Allen P.M., 1990, "Why the Future is not what it was", Futures, July/August p555-570.

Allen P.M. and M.Lesser, 1991, "Evolutionary Human Systems: Learning, Ignorance and Subjectivity", p160-171 in Evolutionary Theories of Economic and Technological Change, Eds Saviotti and Metcalfe, Harwood Chur, Switzerland.

Allen P.M. and J.M.McGlade, 1986, "Dynamics of Discovery and Exploitation: the Scotian Shelf Fisheries", Can.J. of Fish. and Aquat. Sci. Vol.43, N0.6

Allen P.M. and J.M.McGlade, 1987a, "Evolutionary Drive: The Effect of Microscopic Diversity, Error Making and Noise", Foundations of Physics, Vol.17, No.7, July.

Allen P.M. and J.M.McGlade, 1987b, "Managing Complexity: a Fisheries Example", Report to the United Nations University, Tokyo.

Allen P.M. and J.M.McGlade, 1989, "Optimality, Adequacy and the Evolution of Complexity", in "Structure, Coherence and Chaos in Dynamical Systems", Eds. Christiansen and Parmentier, Manchester University Press, Manchester.

Allen P.M. and M. Sanglier,1978 "Dynamic Models of Urban Growth", J.Social and Biological Structures, 1, pp265-280

Allen P.M. and M. Sanglier, 1979, "ADynamic Model of Growth in a Central Place System", Geographical Analysis, Vol.11, No 3. p258-272

Allen P.M. and M. Sanglier, 1981, Urban Evolution, Self-Organization and Decision Making", Environment and Planning A, p167-183.

Arrow K. and G.Debreu,1954, "Existence of an equilibrium for a competitive economy", Econometrica.

Arthur B., 1988, "Self-Reinforcing Mechanisms in Economics", in The Economy as an Evolutionary Complex System, Eds Anderson, Arrow and Pines, Addison-Wesley, New York

Blank, S.C., 1991, "Chaos" in Futures Markets ? A Nonlinear Dynamical Analysis", Journal of Futures Markets, 11:p711-728.

Blume, M.; Stanbaugh, R., 1983, "Biases in Computed Returns: An Application to the Size Effect", Journal of Financial Economics, 12:p33-56.

Chen, P., 1988, "Empirical and Theoretical Evidence of Economic Chaos", System Dynamics Review 4.

Clark W.C. and R.E. Munn, 1986, "Sustainable Development for the Biosphere", Cambridge University Press, Cambridge.

Day, R., 1983, "The Emergence of Chaos from Classical Economic Growth", Quarterly Journal of Economics, 98:201-213.

Debondt, W.; Thaler, R., 1987, "Further Evidence on Investor Overreaction and Stock Market Seasonality", Journal of Finance, 42:p557-581.

Fama, E.F., 1970, "Efficient Capital Markets: A Review of Theory and Empirical Work", Journal of Finance, 25:p383-423.

Grandmont, J.; Malgrange, P., 1986, "Nonlinear Economic Dynamics : Introduction", Journal of Economic Theory 40.

Gustave, Le Bon, 1922, "Paychologie des Foules". Reprinted as "The Crowd", Macmillan, New York.

Hsieh, D.A., 1991, "Chaos and Nonlinear Dynamics : Application to Financial Markets", The Journal of Finance, 66:p1639-1877.

Janis, Irving, 1972, "Victims of GroupThink", Houghton Mifflin, Boston.

Kesley, D., 1988, "The Economics of Chaos or the Chaos of Economics", Oxford Economic Papers 40.

Larrain, M., 1991, "Empirical Tests of Chaotic Behaviour in a Nonlinear Interest Rate Model", Financial Analysts Journal, Sept-Oct, p51-62.

Murphy, John J., 1986, "Techncial Analysis of the Futures Market", New York Institute of Finance, New York.

Muth, J.F., 1961, "Rational Expectations and the Theory of Price Movements", Econometrica, 29:p315-335.

Nicolis G. and I.Prigogine, 1977, "Self-Organization in Non-Equilibrium Systems", Wiley Interscience, New York.

Osborne, M.F.M., 1964, "Brownian Motion in the Stock Market" in P. Cootner, (ed.), "The Random Character of Stock Market Prices", Cambridge, MA:M.I.T. Press. (Originally published in 1959)

Peters, E., 1991a, "A Chaotic Attractor for the S & P 500", Financial Analysts Journal.

Peters, E., 1989, "Fractal Structure in the Capital Markets", Financial Analysts Journal.

Peters, E., 1991b, "R/S Analysis using Logarithmic Returns : A Technical Note", Financial Analysts Journal.

Prigogine I. and I.Stengers, 1987, "Order out of Chaos", Bantam Books, New York.

Rozeff, M.S.; Kinney, W.R., 1976, "Capital Market Seasonality: The Case of Stock Returns"; Journal of Financial Economics, 3:p379-402.

Savit, R., 1988, "When Random is Not Random : An Introduction to Chaos in Market Prices", The Journal of Futures Markets, 8:p271-289.

Savit, R., 1989, "Nonlinearities and Chaotic Effects in Options Prices", The Journal of Futures Markets, 9:p507-518.

Scheinkman, J.; LeBaron, B., 1989, "Nonlinear Dynamics and Stock Returns", Journal of Business, 62:p311-337.

Soros, George, 1988, "The Alchemy of Finance", Weidenfeld and Nicolson, London.

Wachtel, S.B., 1942, "Certain Observations On Seasonal Movements in Stock Prices", Journal of Business, 15:p184-193.

Philosophical Foundations of Nonlinear Complex Systems

K. Mainzer

Lehrstuhl für Philosophie und Wissenschaftstheorie,
Universität Augsburg, Universitätsstraße 10,
D-8900 Augsburg, Fed. Rep. of Germany

Abstract: The main problems of mankind have become global, complex, and nonlinear. Local changes in the ecological, economic or political system can cause a global crisis. Nonlinear interactions in complex networks have synergetic effects which can neither be traced back to single causes nor be forecast in the long run. Linear thinking and believing that the whole is only the sum of its parts has become dangerous. Individual responsibility for synergetic effects seems to be doubtful or at least questionable. Our ecological problems, for instance, are not caused by some bad individuals, but by the demand for welfare and economic growth of mankind. Thus we need new strategies to deal with nonlinear complex systems and to evaluate our actions ethically. Synergetics promises to deliver problem-solving procedures founded by the physical, chemical, and biological evolution of nature. In philosophy we have to analyze the conceptual foundations of this approach in order to estimate its interdisciplinary applications not only in the natural sciences but in the humanities, too. We begin with some methodological remarks on the principles of synergetics. Then the evolution of matter, life, human society, and the evolution of natural and artificial intelligence is discussed. Our philosophical aim are the 'prolegomena' of an epistemology and ethics mastering the problems of a nonlinear complex reality.

1 From Linear to Nonlinear Causality

In the history of science the concepts of the humanities have often been influenced by physical theories. In the age of mechanization T. Hobbes described the state as a machine ('Leviathan') with its citizens as cog wheels. For Lamettrie the human soul was reduced to the gear drive of an automaton. A. Smith explained the mechanism of market by an 'invisible' force like Newton's gravitation. In classical mechanics causality is deterministic in the sense of the Newtonian or Hamiltonian equation of motion. A conservative system is characterized by its reversibility (i.e. symmetry or invariance) of time and the conservation of energy. Celestial mechanics and the pendulum without friction are prominent examples. Dissipative systems are irreversible, for instance Newton's force with a friction term.

But, in principle, nature was regarded as a huge conservative and deterministic system the causal events of which can be forecast and traced back for each point of time in future and past if the initial state is well-known ('Laplacian demon'). It was H. Poincaré who recognized that celestial mechanics is no completely calculable clockwork even with the restrictions of conservation and determinism. The causal interactions of all planets, stars, and celestial bodies are nonlinear in the sense that their mutual effects

Springer Series in Synergetics, Vol. 62 **Interdisciplinary Approaches to Nonlinear Complex Systems** – Eds.: H. Haken and A. Mikhailov

can lead to chaotic trajectories ('3-body-problem'). Nearly sixty years after Poincaré's discovery A.N. Kolmogorov (1954), V.I. Arnold (1963), and J.K. Moser proved the so-called KAM-theorem: Trajectories in the phase space of classical mechanics are neither completely regular nor completely irregular, but they depend very sensitively on the chosen initial states. Tiny fluctuations can cause chaotic developments ('butterfly effect') [1].

In this century quantum mechanics has become the fundamental theory of physics. In Schrödinger's wave mechanics the quantum world is believed to be conservative and linear. In the 1st quantisation classical systems described by a Hamilton-function are replaced by quantum systems (for instance electrons or photons) described by a Hamilton-operator. These systems are assumed to be conservative, i.e. non-dissipative, invariant with respect to time reversion and thus satisfying the conservation law of energy. States of a quantum system are described by vectors ('wave-functions') of a Hilbert-space spanned by the eigenvectors of its Hamilton-operator. The causal dynamics of quantum states is determined by a deterministic differential equation ('Schrödinger equation') which is linear in the sense of the superposition principle, i.e. solutions of this equation ('wave functions' or 'state vectors') can be superposed like in classical optics. The superposition or linearity principle of quantum mechanics delivers correlated ('entangled') states of combined systems which are highly confirmed by the EPR-experiments (A. Aspect 1981). In an entangled pure quantum state of superposition an observable can only have indefinite eigenvalues. It follows that the entangled state of a quantum system and a measuring apparatus can only have indefinite eigenvalues. But in the laboratory the measuring apparatus shows definite measurement values. Thus, the linear quantum dynamics cannot explain the measurement process [2].

In the Copenhagen interpretation of Bohr, Heisenberg, etc. the measurement process is explained by the so-called 'collapse of the wave-packet', i.e. splitting up of the superposition state into two separated states of measurement apparatus and measured quantum system with definite eigenvalues. Obviously, we must distinguish the linear dynamics of quantum systems from the nonlinear act of measurement. This non-linearity in the world is sometimes explained by the emergence of human consciousness. E.P. Wigner (1961) suggested that the linearity of Schrödinger's equation might fail for conscious observers, and be replaced by some nonlinear procedure according to which either one or the other alternative would be resolved out. But Wigner's interpretation forces us to believe that the linear quantum superpositions would be resolved into separated parts only in those corners of the universe where human or human-like consciousness emerges. In the history of science anthropic or teleological arguments often showed that there were weaknesses and failures of explanation in science. Thus, some scientists like R. Penrose suppose that the linear dynamics of quantum mechanics is not convenient to explain cosmic evolution with the emergence of consciousness [3]. He argues that a unified theory of linear quantum mechanics and nonlinear general relativity could at least explain the separated states of macroscopic systems in the world. A measuring apparatus is a macroscopic system, and the measurement process is irreversible far from thermal equilibrium. Thus an explanation could only succeed in a unified nonlinear theory.

Even the generalization of Schrödinger's wave mechanics to quantum field theory is already nonlinear. In quantum field theory field functions are replaced by field operators in the so-called 2nd quantisation. The quantum field equation with a 2-particle potential, for instance, contains a nonlinear term corresponding to pair creation of elementary particles. In general the reactions of elementary particles in quantum field theory are essentially nonlinear phenomena. The interactions of an elementary particle cause its quantum states to have only a finite duration and thereby to violate the reversibility of time. Thus even the quantum world itself is neither conservative nor linear in general [4].

2 Synergetics and the Evolution of Matter

Complexity is an essential property of physical reality besides nonlinearity. All macroscopic systems like stones or planets, clouds or fluids, plants or animals, animal populations or human societies consist of component elements like atoms, molecules, cells or organisms. The behaviour of single elements in complex systems with huge degrees of freedom can neither be forecast nor traced back. The deterministic description of single elements must be replaced by the evolution of probabilistic distributions.

Since the presocratics it has been a fundamental problem of natural philosophy to discover how order arises from complex, irregular, and chaotic states of matter. Heraclitus believed in an ordering force of energy ('logos') harmonizing irregular interactions and creating order states of matter. Modern thermodynamics describes the emergence of order by the mathematical concepts of statistical mechanics. We distinguish two kinds of phase transition ('self-organization') for ordered states: The conservative self-organization means the phase transition of reversible structures in thermal equilibrium. Typical examples are the growth of snow crystals or the emergence of magnetisation in a ferromagnet by annealing the system to a critical value of temperature. Conservative selforganization mainly creates ordered structures with low energy at low temperatures, which are described by a Boltzmann distribution. Dissipative selforganization means the phase transition of irreversible structures far from thermal equilibrium. Macroscopic patterns arise from the complex nonlinear cooperation of microscopic elements when the energetic interaction of the dissipative ('open') system with its environment reaches some critical value. Philosophically speaking the stability of the emergent structures is guaranteed by some balance of nonlinearity and dissipation. Too much nonlinear interaction or dissipation would destroy the structure.

As the conditions of dissipative phase transition are very general, there is a broad variety of interdisciplinary applications. A typical physical example is the laser. In chemistry, the concentric rings or moving spirals in the Belousov-Zhabotinski (BZ) reaction arise when specific chemicals are poured together with a critical value. The competition of the separated ring waves show the nonlinearity of these phenomena very clearly, because in the case of a superposition principle the ring waves would penetrate each other like optical waves.

The phase transitions of nonlinear dissipative complex systems are explained by synergetics [5]. In a more qualitative way we may say that old structures become unstable and break down by changing control parameters. On the microscopic level the stable modes of the old states are dominated by unstable modes (Haken's 'slaving principle'). They determine order parameters which describe the macroscopic structure and patterns of systems. There are different final patterns of phase transitions corresponding to different attractors. A survey is given by different attractors of a stream, the velocity of which is accelerated step by step. At a first level a homogeneous state of equilibrium is shown ('fixed point'). At a higher level of velocity the bifurcation of two or more vortices can be observed corresponding to periodic and quasi-periodic attractors. Finally the order decays into deterministic chaos as a fractal attractor of complex systems. Philosophically I want to underline that in synergetics the microscopic description of matter is distinguished from the macroscopic order states. Thus the synergetic concept of order reminds me of Heraclitus' 'logos' or Aristotle's 'form' which produces the ordered states of nature in a transformative process of matter. But, of course, in antiquity a mathematical description was excluded.

In a more mathematical way the microscopic view of a complex system is described by the evolution equation of a state vector where each component depends on space and time and where the components may mean the velocity components of a fluid, its temperature field, or in the case of chemical reactions, concentrations of chemicals. The slaving principle of synergetics allows us to eliminate the degrees of freedom which refer to the stable modes. In the leading approximation the evolution equation can be transformed into a specific form for the nonlinearity which applies to those systems where a competition between patterns occurs. The amplitudes of the leading terms of unstable modes are called order parameters. Their evolution equation describes the emergence of macroscopic patterns. The final patterns ('attractors') are reached by a transition which can be understood as a kind of symmetry breaking. Philosophically speaking the evolution of matter is caused by a process of symmetry breaking which was earlier mentioned by Heraclitus [6].

3 Synergetics and the Evolution of Life

In the history of science and philosophy people believed in a sharp difference of 'dead' and 'living' matter. Aristotle interpreted life as a power of selforganization ('entelechy') driving the growth of plants and animals to their final form. A 'living' system is able to move by itself, while a 'dead' system can only be moved from outside. Life was explained by teleology, i.e. by non-causal ('vital') forces aiming at some goals in nature. In the 18th century Kant showed that selforganization of living organisms cannot be explained by a mechanical system of Newtonian physics. In a famous quotation he said that the 'Newton for explaining a blade of grass is still lacking'. In the 19th century the 2nd law of thermodynamics describes the irreversible process of closed systems to a state of maximal entropy or disorder. But how can one explain the emergence of order in Darwin's evolution of life? Boltzmann stressed that living organisms are open dissipative systems in exchange with their environment which do not violate the 2nd law of closed systems. But nevertheless in the statistical interpretation from Boltzmann to Monod the emergence of life can only be a contingent event, a local cosmic fluctuation 'at the boundary of universe' [7].

In the framework of synergetics the emergence of life is not contingent, but necessary and lawful in the sense of dissipative selforganization. Only the conditions for the emergence of life (for instance on the planet earth) may be contingent in the universe. In general, biology distinguishes the ontogenesis, i.e. the growth of organisms, and the phylogenesis, i.e. the evolution of species. In any case we have complex dissipative systems the development of which can be explained by the evolution of (macroscopic) order parameters caused by nonlinear (microscopic) interactions of molecules, cells etc. in phase transition far from thermal equilibrium [8]. Forms of biological systems (plants, animals, etc.) are described by order parameters. Aristotle's teleology of goals in nature is interpreted as attractors in phase transitions. But no special 'vital' or 'teleological' forces are necessary. Philosophically, the emergence of life can be explained in the framework of nonlinear causality and dissipative selforganization, although it may be described in a teleological language by heuristic reasons.

I remind the reader of the prebiological evolution of biomolecules which was analyzed and simulated by M. Eigen et al. Spencer's idea that the evolution of life is characterized by increasing of complexity can be made precise in the context of dissipative selforganization. It is well-known that Turing analyzed a mathematical model of organisms by complex cellular systems. Gerisch, Meinhardt, et al. described the growth of an organism (e.g. a slime mould) by evolution equations for the aggregation of cells. The nonlinear interactions of the amoebae cause the emergence of an macroscopic organism like the slime mould when some critical value of cellular nutrition in the environment is reached. The evolution of the order parameter corresponds to the aggregation forms during the phase transition of the macroscopic organism. The mature multicellular body can be interpreted as the 'goal' or (better) 'attractor' of organic growth.

Even the ecological growth of biological populations may be simulated by the concept of synergetics. Ecological systems are complex dissipative systems of plants or animals with mutual nonlinear interactions and metabolism with its environment. The symbiosis of two populations with their source of nutrition can be described by three coupled differential equations which were already used by E.N. Lorenz to describe the development of weather in meteorology. In the 19th century the Italian mathematicians Lotka und Volterra described the development of two populations in ecological competition. The nonlinear interactions of the two complex populations are determined by two coupled differential equations of prey and predator fishes. The evolution of the coupled systems have stationary points of equilibrium. The attractors of evolution are periodic oscillations (limit cycles).

Synergetics allows us to analyze the nonlinear causality of ecological systems in nature. Since the industrial revolution human society has become more and more involved in the ecological cycles of nature. But the complex balance of natural equilibria is highly endangered by the linear way of traditional industrial productions. People assumed that nature contains endless sources of energy, water, air, etc. which can be used without disturbing the natural balance. They produce an endless mass of goods without considering their synergetic effects like the ozone hole or waste utilization. The evolution of life is transformed into the evolution of human society.

4 Synergetics and the Evolution of Human Society

In the humanities one usually strictly distinguishes between biological evolution and the history of human society. The reason is that the development of nations, markets, and cultures is assumed to be guided by the intentional behaviour of humans, i.e. human decisions based on intentions, values, etc. From a microscopic view-point we may, of course, observe single individuals with their intentions, beliefs, etc. But from a macroscopic view the development of nations, markets, and cultures is not only the sum of its parts. Mono-causality in politics and history is, as we all know, a false and dangerous way of linear thinking. Synergetics seems to be a successful strategy to handle complex systems even in the humanities. Obviously it is not necessary to reduce cultural history to biological evolution in order to apply synergetics interdisciplinarily. Contrary to any reductionistic kind of naturalism and physicalism we recognize the characteristic intentional features of human societies. Thus synergetics may be a method of bridging the gap between the natural sciences and the humanities that was criticized in Snow's famous 'two cultures'.

In the framework of synergetics the behaviour of human populations is explained by the evolution of (macroscopic) order parameters which is caused by nonlinear (microscopic) interactions of humans or human subgroups (states, institutions, etc.). Social or economic order is interpreted as attractors of phase transitions. Allen et al. analyze the growth of urban regions. From a microscopic view-point the evolution of populations in single urban regions is mathematically described by coupled differential equations with terms and functions refering to the capacity, economical production, etc. of each region. The macroscopic development of the whole system is illustrated by computer-assisted graphics with changing centres of industrialization, recreation, etc. which are caused by nonlinear interactions of single urban regions (for instance advantages and disadvantages of far and near connections of transport, communication, etc.). An essential result of the synergetic model is that the urban development cannot be explained by the free will of single persons. Although people of local regions are acting with their individual intentions, plans, etc., the tendency of the global development is the result of nonlinear interactions.

Another example of interdisciplinary application of synergetics is Weidlich's model of migration [9]. He distinguishes the micro-level of individual decisions and the macro-level of dynamical collective processes in the society. The probabilistic macro-processes with stochastic fluctuations are described by the master equation of human socioconfigurations. Each component of a socioconfiguration refers to a subpopulation with a characteristic vector of behaviour. The macroscopic development of migration in a society could be illustrated by computer-assisted graphics with changing centres of mixtures, ghettos, wandering, and chaos which are caused by nonlinear interactions of social subpopulations. The differences between human and non-human complex systems are obvious in this model. On the microscopic level human migration is intentional (i.e. guided by considerations of utility) and nonlinear (i.e. depending on individual and collective interactions). A main result of synergetics is again that the effects of national and international migration cannot be explained by the free will of single persons. I think migration is a very dramatic topic of today demonstrating how dangerous linear and mono-causal thinking may be. It is not sufficient to have good intentions without considering the nonlinear effects of single decisions. Linear thinking and acting may provoke global chaos, although we locally act with the best intentions.

It is a pity to say that in economics linear models are still dominant. From a qualitative point of view A. Smith's model of a free market can already be explained by selforganization. Smith underlined that the good or bad intentions of individuals are not essential. Contrary to a centralized economical system the equilibrium of supply and demand is not directed by a program-controlled central processor, but the effect of an 'invisible hand' (Smith), i.e. nothing other than the nonlinear interaction of consumers and producers. The recent interest of economists in nonlinear dissipative systems is inspired by the growth of knowledge-based high-tech industries with positive feedback (i.e. increasing produce depending on increasing know-how like electronics, computer industries, etc.) in contrast to the old industries with negative feedback (i.e. decreasing produce depending on limited ressources like coal or steel). In general economic processes are very complex and demand nonlinear dissipative models. Recall the different attractors from economical cycles to financial chaos which can only be explained as synergetic effects by nonlinear interactions of consumers and producers, fiscal policy, stock market, unemployment, etc. [10]. Even in management policy synergetic models are discussed in order to support creativity and innovation by nonlinear cooperation of all levels of management and production. The synergetic analysis shows that economic processes are to be embedded in the ecological cycles of nature. It must be the main intention of our politics to achieve a nonlinear complex system of economics and ecology that maintains a balance between human society and nature.

5 Synergetics and the Evolution of Natural and Artificial Intelligence

The most speculative interdisciplinary application of synergetics is the evolution of natural and artificial intelligence (AI). In the history of philosophy and science there have been many different suggestions of solutions concerning the mind-body problem. Materialistic philosophers like Democritus, Lamettrie, etc. intended to reduce mind to atomic interactions. Idealists like Plato, Penrose, etc. emphasized that mind is completely independent of matter and brain. For Descartes, Eccles, etc. mind and matter are separate substances interacting with each other. Leibniz believed in a metaphysical parallelism of mind and matter because they cannot interact physically. According to Leibniz mind and matter are supposed to exist in 'pre-established harmony' like two synchronized clocks. Modern philosophers of mind like Searle defend a kind of evolutionary naturalism. Searle argues that mind is characterized by intentional mental states which are intrinsic features of the human brain's biochemistry and which therefore cannot be simulated by computers.

But synergetics cannot be reduced to these more or less one-sided positions. Synergetics is an interdisciplinary methodology to deal with nonlinear complex systems like the cellular organ of brain. The emergence of mental states (for instance pattern recognition, feelings, thoughts) is explained by the evolution of (macroscopic) order parameters of cerebral assemblies which are caused by nonlinear (microscopic) interactions of neural cells in learning strategies far from thermal equilibrium. Cell assemblies with mental states are interpreted as attractors (fixed points, periodic, quasi-periodic, chaotic) of phase transitions [11].

If the brain is regarded as a complex system of neural cells, then its dynamics is assumed to be described by the nonlinear mathematics of neural networks. Pattern recognition, for instance, is interpreted as a kind of phase transition in analogy to the evolution equations which are used for pattern emergence in physics, chemistry, and biology. Philosophically we get an interdisciplinary research program that should allow us to explain neurocomputational selforganization as a natural consequence of physical, chemical, and neurobiological evolution by common principles. As in the case of pattern formation a specific pattern of recognition (for instance a prototype face) is described by order parameters to which a specific set of features belongs. Once some of the features which belong to the order parameter are given (for instance a part of a face), the order parameter will complement these with the other features so that the whole system acts as associative memory (for instance the reconstruction of a stored prototype face from an initially given part of that face). According to Haken's slaving principle the features of a recognized pattern correspond to the enslaved subsystems during pattern formation. The order parameter equations allow a new kind of (non-Hebbian) learning, namely a strategy to minimize the number of synapses. In contrast to neurocomputers of the spin-glass type (for instance Hopfield systems), the neurons are not threshold elements but rather perform simple algebraic manipulations like multiplication and addition. Besides deterministic homogeneous Hopfield networks there are so-called Boltzmann machines with a stochastic network architecture of non-deterministic processor elements and a distributed knowledge representation which is described mathematically by an energy function. While Hopfield systems use a Hebbian learning strategy, Boltzmann machines favour a backpropagation strategy (Widrow-Hoff rule) with hidden neurons in a many-layered network. In general it is the aim of a learning algorithm to diminish the information-theoretic measure of the discrepancy between the brain's internal model of the world and the real environment via selforganization [12].

The recent revival of interest in the field of neural networks is mainly inspired by the successful technical applications of statistical mechanics and nonlinear dynamics to solid state physics, spin glass physics, chemical parallel computers, optical parallel computers, and – in the case of synergetic computers – to laser systems. Other reasons are the recent development of computing resources and the level of technology which make a computational treatment of nonlinear systems more and more feasible. Philosophically, traditional topics of epistemology like perception, imagination, and recognition may be discussed in the interdisciplinary framework of synergetics [13].

But what about the emergence of consciousness, self-consciousness, and intentionality? In synergetics we have to distinguish between external and internal states of the brain. In external states of perception and recognition order parameters correspond to neural cell assemblies representing patterns of the external world. Internal states of the brain are nothing other than self-referential states, i.e. mental states refering to mental states and not to external states of the world. In the traditional language of philosophy we say that humans are able to reflect on themselves ('self-reflection') and to refer external situations of the world to their own internal state of feeling and intentions (intentionality). In recent inquiries of neurobiology scientists speculate that the emergence of consciousness and selfconsciousness depends on a critical value of the production rate for 'meta-cell-assemblies', i.e. cell-assemblies representing cell-assemblies which are again representing cell-assemblies, etc. as neural realization of self-reflection. But this hypothesis (if successful) could only explain the structure of emergent features like consciousness. Of course, mathematical evolution equations of cell assemblies do not enable us to feel like our neighbour. In this sense - that is the negative message - science is blind. But otherwise - that is the positive message - personal subjectivity is saved [14].

Anyway the synergetic approach solves an old metaphysical puzzle which was described by Leibniz in the following picture: If we imagine the brain as a big machine which we may enter like the internal machinery of a mill, we shall only find its single parts like the cog wheels of the mill and never the mind, not to mention the human soul. Of course, on the microscopic level we can only describe the development of neurons as cerebral parts of the brain. But, on the macroscopic level, the nonlinear interactions in the complex neural system cause the emergence of cell assemblies refering to order parameters which cannot be identified with the states of single cerebral cells. The whole is not the sum of its parts. Concerning the distinction of so-called natural and artificial intelligence it is important to see that the principles of synergetics do not depend on the biochemistry of human brain. Human brain is a 'natural' model of synergetic principles in the sense that the cerebral complex system is a product of physical and biological evolution. But other ('artificial') models produced by human technology are possible, although there will be technical and ethical limits to their realization.

Summary

Synergetics is evidently a successful strategy to handle nonlinear complex systems. We have discussed examples of applications from quantum physics, hydrodynamics, chemistry, and biology to economics, sociology, neurology, and AI. What is the reason behind the successful applications in the natural sciences and humanities? Synergetics is not reduced to special natural laws of physics, although its mathematical principles were discovered and at first successfully applied in physics (to the laser). Thus it is an interdisciplinary methodology to explain the emergence of certain macroscopic phenomena via the nonlinear interactions of microscopic elements in complex systems. Macroscopic phenomena may be forms of light waves, fluids, clouds, chemical waves, biomolecules, plants, animals, populations, markets, and cerebral cell assemblies which are characterized by order parameters (Table 1).

Philosophically it is important to see that order parameters are not be reduced to the microscopic level of atoms, molecules, cells, organisms, etc. of complex systems. In some cases they are measurable quantities (for instance the field potential of a laser). In other cases they are qualitative properties (for instance geometrical forms of patterns). Nevertheless order parameters are not mere theoretical concepts of mathematics without any reference to reality. Actually they represent properties of real macroscopic phenomena like for instance field potentials, social or economical power, feelings or even thoughts. Who will deny that feelings and thoughts can change the world? But synergetics is not a metaphysical process ontology. The synergetic principles (among others) deliver an heuristic scheme to construct models of nonlinear complex systems in the natural sciences and the humanities. If these models can be mathematized and their poperties quantified, then we get empirical models which may or may not fit the data. The slaving principle shows another advantage. As it diminishes the high number of degrees of freedom in a complex system, synergetics is not only heuristic, mathematical, empirical and testable, but economical too. Namely, it satisfies the famous principle of Ockham's razor which tells us to cut up superfluous entities.

Synergetics suggests that physical, social, and mental reality is nonlinear and complex. This essential result of synergetic epistemology demands severe consequences of our behaviour. As we underlined in earlier chapters, linear thinking may be dangerous in a nonlinear complex reality. Recall, as one example, the demand for a well-balanced complex system of ecology and economics. Our physicians and psychologists must learn to consider humans as complex nonlinear entities of mind and body. Linear thinking may fail to yield a successful diagnosis. Local, isolated, and 'linear' therapies of medical treat-

Table 1. Selforganization of nonlinear complex systems

DISCIPLINE	SYSTEM	ELEMENTS	SELFORGANISATION	ORDER PARAMETER
quantum physics	laser	atoms (photons)	phase transition	form of light waves
hydrodynamics	fluids	molecules	phase transition	form of fluids
meteorology	weather	molecules	phase transition	form of clouds
geology	lava	molecules	phase transition	hexagonal form (Bénard cells)
chemistry	BZ-reaction	molecules	phase transition	forms of spiral or rings (chemical waves)
biology	biomolecules	molecules	phase transition	structural form
	organisms	cells	organic growth	organic forms (plants, animals)
	population	organisms	evolution of populations	form of population (interactional form)
economics	economic systems	consumers, producers etc.	mechanism of market (e.g. supply and demand)	form of market (interactional form)
sociology	societies	humans, institutions etc.	history	interactional form
neurology (psychology)	brain	neurons	recognition (learning)	forms of neural cell assemblies representing external or internal ('self-referential') states
artificial intelligence (AI)	neural AI-networks	AI-neurons	learning algorithms	forms of neural AI-cell assemblies representing external or internal ('self-referential') states

ment may cause negative synergetic effects. In politics and history, we must remember that mono-causality may lead to dogmatism, intolerance, and fanatism. As the ecological, economical, and political problems of mankind have become global, complex, and nonlinear, the traditional concept of individual responsibility is questionable. We need new models of collective behaviour depending on the different degrees of our individual faculties and insights. In short: Synergetics demands new consequences in epistemology and ethics. Finally, it offers a chance to prevent chaos in a nonlinear complex world and to utilize the creative possibilities of synergetic effects.

Acknowledgement

This work was partially supported by a DFG-project on 'Computer, Chaos und Selbstorganisation' (Ma 842/4-1).

References

1. K. Mainzer, W. Schirmacher (eds.): Quanten, Chaos und Dämonen. Erkenntnistheoretische Aspekte der modernen Physik, B.I.-Wissenschaftsverlag, Mannheim, Leipzig, Wien, Zürich 1993.
2. J. Audretsch, K. Mainzer (eds.): Wieviele Leben hat Schrödingers Katze? Zur Physik und Philosophie der Quantenmechanik, B.I.-Wissenschaftsverlag, Mannheim, Wien, Zürich 1990.
3. R. Penrose: The Emperor's New Mind. Concerning Computers, Minds, and the Laws of Physics, Oxford University Press, Oxford, New York, Melbourne 1989.
4. C.P. Enz: 'Beschreibung nicht-konservativer nicht-linearer Systeme I-II', Physik in unserer Zeit **4** 119-126 (1979) (I), **5** 141-144 (1979) (II).
5. H. Haken: Synergetics. An Introduction, 3rd Edn. Springer, New York, Heidelberg, Berlin, Tokyo 1983, German translation ('Synergetik'), 3rd Edn. Springer, Heidelberg 1983.
6. K. Mainzer: Symmetrien der Natur. Ein Handbuch zur Natur- und Wissenschaftsphilosophie, De Gruyter, Berlin, New York 1988, English translation ('Symmetries of Nature') De Gruyter, New York 1993.
7. E.P. Fischer, K. Mainzer (eds.): Was ist Leben?, Piper, München, Zürich 1990.
8. H. Haken, M. Haken-Krell: Entstehung von biologischer Information und Ordnung, Wissenschaftliche Buchgesellschaft, Darmstadt 1989.
9. Compare [1].
10. W.A. Barnett, J. Geweka, K. Shell (eds.): Economics Complexity: Chaos, Sunspots, Bubbles, and Nonlinearity, Cambridge University Press, New York, Port Chester, Melbourne, Sydney 1989;
    K. Mainzer (ed.): Ökonomie und Ökologie/Économie et Écologie, Paul Haupt, Bern, Zürich 1992.
11. H. Haken, M. Stadler (eds.): Synergetics of Cognition, Springer, Berlin, Heidelberg, New York 1989;
    H. Haken, M. Haken-Krell: Erfolgsgeheimnisse der Wahrnehmung. Synergetik als Schlüssel zum Gehirn, Deutsche Verlags-Anstalt, Stuttgart 1992;
    W. Singer: 'Zur Selbstorganisation kognitiver Strukturen', in Gehirn und Bewußtsein, ed. by E. Pöppel (VCH Weinheim 1989) pp. 45–59.
12. H. Haken (ed.): Neural and Synergetic Computers, Springer, Berlin, Heidelberg, New York 1988.

13. K. Mainzer: 'Philosophical Concepts of Computational Neuroscience', in Parallel Processing in Neural Systems and Computers (North-Holland, Amsterdam, New York, Oxford, Tokyo 1990), ed. by R. Eckmiller, G. Hartmann, G. Hauske, pp. 9–12; 'Knowledge-based Systems. Remarks on Philosophy of Technology and Artificial Intelligence', Journal for General Philosophy of Science 1 47–74 (1990); 'Die Evolution intelligenter Systeme', Zeitschrift für Semiotik 12 81–104 (1990).
14. K. Mainzer: 'Künstliche Intelligenz, Neuroinformatik und die Aufgabe der Philosophie', Forum für interdisziplinäre Forschung 1 56–62 (1992).

Diversity and Collective Action

B. A. Huberman and N. S. Glance

Dynamics of Computation Group, Xerox Palo Alto Research Center,
Palo Alto, CA 94304, USA

Abstract

We elucidate the dynamics of ongoing collective action among intentional agents with diverse beliefs and imperfect information. Their decisions on whether or not to contribute to the collective good depend not only on the past but also on their expectations as to how their actions will affect those of others. We show that in attempts at collective action the onset of overall cooperation can take place in a sudden and unexpected way. Likewise, defection can appear out of nowhere in very large, previously cooperating groups. These outbreaks mark the end of long transient states in which defection or cooperation persists in groups that cannot sustain it indefinitely. Computer experiments demonstrate these predictions, as well as verifying that diversity of beliefs among individuals acts as an additional source of uncertainty, instigating the outbreaks.

1 Introduction

Collective action problems create difficult quandaries for societies, be they social, economic, or organizational. Whenever a group of intentional agents collaborates in a collective task that produces some overall utility to the group, the potential for a dilemma arises. In cases where the gain to an individual for collaborating is less then its cost for participating, it may rationally choose not to cooperate and instead to free ride on the effort of others. When this logic holds for all individuals, no sustained cooperation ensues. The problem occurs when the benefit to be accrued by overall cooperation offsets individual costs, since rationality on the part of each individual leads to failure in achieving a collective good beneficial to all. The resolution of such dilemmas, which are the subject of study in the social sciences [1–6], underlies successful attempts at ongoing collective action, such as the functioning of large organizations, the adoption of new technologies [7], and the mobilization of political movements [8].

In human societies, an essential element contributing to the likelihood of collective action is that individuals differ from each other both in their beliefs and in their estimates of the costs and benefits of contributing to the collective good. Thus, whereas one individual might decide to participate in the group effort on the basis of a few others having joined, another might wait until many others have already done so. This diversity comes into play in many instances of collective action.

In studying collective action one may then ask the following question: if agents make decisions on whether or not to cooperate on the basis of imperfect information

Springer Series in Synergetics, Vol. 62 **Interdisciplinary Approaches to Nonlinear Complex Systems** – Eds.: H. Haken and A. Mikhailov

about the group activity, and incorporate expectations on how their decision will affect other agents, can overall cooperation be sustained for long periods of time? Moreover, how do expectations, diversity of beliefs, and group size affect cooperation?

Collective action problems are characterized by the impossibility of exclusion; that is, no member of a group engaged in collective action can be excluded from enjoying the benefits of the group's efforts. On the other hand, collective goods can have varying amounts of jointness of supply, which is the degree to which one agent's consumption of the good does not reduce the amount available to any other. The amount of jointness determines in part the dependence of cooperative outcomes on the size of the group [5]. Thus, for example, some studies have shown that overall cooperation is undermined as the group increases in size [1, 9], others that, to the contrary, that it is more likely for larger groups [8]. The latter result was obtained for public goods, a special subset exhibiting perfect jointness of supply, while the former holds for divisible goods, those with limited jointness of supply. In this paper, we consider divisible collective goods.

Regardless of whether or not a group of a given size can exhibit ongoing cooperation, there remains the issue of how is it that such a state is reached, if ever. For one can imagine situations whereby a group initially exhibits cooperative behavior in spite of its being too large to do so, only to gradually evolve into collective defection. Conversely, a large non-cooperative group could undergo a drastic reduction in size and the relevant question then becomes how long before the switch to global cooperation occurs.

In order to answer these questions, we present and study a dynamical model of ongoing collective action among intentional agents whose choices depend not only on the past but also on their expectations as to how their actions will affect those of others. In this model agents act on the basis of information that can be uncertain at times. We show that under these conditions the onset of overall cooperation can take place in a sudden and unexpected way. Likewise, defection can appear out of nowhere in very large, previously cooperating groups. These outbreaks mark the end of long transient states in which defection or cooperation persists in groups that cannot sustain it indefinitely. Moreover, we study the effects of diversity on these phenomena. First of all, we find that diversity acts as an additional source of uncertainty, thus shortening the time to an outbreak. Also, we show that when several subgroups with different beliefs are merged into a larger group, overall cooperation (or defection) can appear in stages.

Following a long tradition [10, 3, 9], in Section 2 we study the free rider problem as a repeated n-person prisoners' dilemma, wherein the benefit obtained by an individual from cooperating in producing the good is outweighed by the cost of cooperation for the one-shot game. We recast the interaction as an asynchronous dynamic game in which each individual reconsiders its decision at an average reevaluation rate α, using delayed information of the level of provision of the good. The iterated game is of finite duration, and individuals decide to cooperate or defect by determining which choice maximizes their *expected* share of the good for the remainder of the game. In addition, uncertainty is introduced into the relation between individual effort and group performance to model imperfect information and bounded rationality. The incorporation of diversity into the model is left until Section 4, so that its effects may be more apparent.

The type of expectations that we consider consist of two components: each individual believes (1) that future aggregate collective behavior is directly influenced by the individual's choices in inverse proportion to the group size; and (2) that the interaction is of finite duration characterized by a horizon length, H. Thus, individuals believe that in the long run their actions encourage similar actions on the part of others in the group. There are various mechanisms by which this might occur: imitation and establishment of conventions or norms are a few. In addition, individuals believe that the ability of their actions to encourage like actions increases with the number of individuals who have chosen to contribute to the collective good (as opposed to free riding). In some sense, this reflects a belief that contributing individuals form a core group whose reactions are much more sensitive to fluctuations in the amount of the good produced.

We show that an individually rational strategy of conditional cooperation emerges from the individuals' expectations and beliefs. Individuals cooperate if they perceive the fraction cooperating to be greater than some critical amount and defect otherwise. This strategy of conditional cooperation is reminiscent of the successful tit-for-tat strategy in 2–player prisoners' dilemma in which a player cooperates if and only if its opponent cooperated in the previous turn and defects otherwise [11].

In Section 3, we analyze the dynamics of fluctuations away from Nash equilibria using a thermodynamic-like formalism [12] for uniform groups without diversity. Besides confirming that there exists a critical group size beyond which cooperation is not sustainable [1, 9], we find additional, intriguing effects. Our results reveal several different dynamical regimes that should be observable as the size of a group changes. In one regime, although cooperation may persist for very long times even for groups exceeding a critical size, group behavior eventually decays to overall defection. In another situation a system can be stuck in a non-cooperative state even though its size is well below that guaranteeing long term cooperation. These effects are shown to depend strongly on the degree of uncertainty pervading the system, as well as on the length of the individual's horizons. To confirm our analytical predictions, we present the results of computer simulations.[1]

In Section 4 we study the effect of diversity among individuals on the phenomena described in Section 3. We find that without changing the critical sizes for cooperation, diversity acts a source of additional uncertainty and shortens the transition time to the long-term stable state of the system. In fact, an analytical form for the increase in uncertainty due to diversity is given. Again, computer simulations are invoked which confirm our predictions. Moreover, these simulations also show that when there is diversity of beliefs among subgroups merged to form a large group, cooperation can be achieved in stages.

Section 5 summarizes our results.

[1] The interested reader is referred Glance and Huberman [13] for additional technical detail beyond that provided in this paper.

2 The Dynamics of Collective Action

The economics of free riding

We consider the ongoing group interaction involved in the production of a non-excludable collective good which exhibits limited jointness of supply. Each individual can either contribute (cooperate) to the production of the good, or not (defect). While no individual can directly observe the effort of another, each member observes instead the collective output and can deduce overall group participation using knowledge of individual and group production functions. We also introduce an amount of uncertainty into the relation between members' efforts and group performance. There are many possible causes for this uncertainty [9]; for example, a member may try but fail to contribute due to unforeseen obstacles. Alternatively, another type of uncertainty might arise due to individuals with bounded rationality occasionally making suboptimal decisions [14, 15]. In any case, we treat here only idiosyncratic disturbances or errors, whose occurrences are purely uncorrelated.

Consequently, we assume that a group member intending to participate does so successfully with probability p. The individual's attempt to cooperate, instead of guaranteeing contribution to the cause, fails with probability $1-p$, with an effect equivalent to a defection. The inverse scenario holds for members intending to defect: an attempt to defect results in zero contribution with probability q, but with probability $1-q$ a defecting member will inadvertently (or suboptimally) contribute to the collective good. Then, as all attempts are uncorrelated, the number of successfully cooperating members, $\widehat{n}_c$, is a mixture of two binomial random variables with mean $<\widehat{n}_c>= pn_c + (1-q)(n-n_c)$, where n_c is the number of members attempting to cooperate within a group of size n.

The limit p and q equal to 1 corresponds to an error-free world of complete information, while p and q equal to 0.5 reflect the case where the effect of an action is completely divorced from intent. Note that whenever p and q deviate from 1, the perceived level of cooperation will differ from the actual attempted amount of participation.

In the simplest case, collective benefits increase linearly in the contributions of the members, at a rate b per cooperating member. Each contributing individual bears a personal cost, c. Let k_i denote whether member i is cooperating (k_i=1) or defecting (k_i=0). Then the utility at time t for member i is

$$U_i(t) = \frac{b}{n}\widehat{n}_c(t) - ck_i. \tag{2.1}$$

Using its knowledge of the functional form of the utility function[2], each individual can deduce the number of individuals effectively cooperating at some time t by inverting Eq. 2.1:

$$\widehat{n}_c(t) = \frac{n}{b}(U_i(t) + ck_i). \tag{2.2}$$

Of course, this estimation will differ from the actual number of individuals intending to cooperate in a manner described by the mixture of two binomial distributions. We also define $\widehat{f}_c(t)$, to denote the fraction, $\widehat{n}_c(t)/n$, of individuals effectively cooperating at time t.

When all members contribute successfully, each receives net benefits $bn/n - c = b - c$. The production of the collective good becomes an n-person prisoners' dilemma when

$$b > c > \frac{b}{n}. \tag{2.3}$$

Thus, although the good of all is maximized when everyone cooperates $(b - c > 0)$, the dominant strategy in the one-shot game is to defect since additional gain of personal participation is less than the private cost $(b/n - c < 0)$.

The logic behind the decision to cooperate or not changes when the interaction is ongoing since future expected utility gains will join present ones in influencing the rational individual's decision to contribute or not to the collective good. In particular, individual expectations concerning the future evolution of the game can play a significant role in each member's decisions. The importance individuals place on the future depends on how long they expect the interaction to last. If they expect the game to end soon, then, rationally, future expected returns should be discounted heavily with respect to known immediate returns. On the other hand, if the interaction is likely to continue for a long time, then members may be wise to discount the future only slightly and make choices that maximize their returns on the long run. Notice that making present choices that depend on the future is rational only if, and to the extent that, a member believes its choices influence the decisions others make.

In the next section we elaborate on one self-consistent set of beliefs that permits individuals engaged in ongoing collective action to make the decision whether or not to contribute.

Expectations

The time scale of the interaction is set by the rate, α, at which members of the group reexamine their choices. Information about the level of cooperation is deduced from

[2] For a justification of the form of the individual utility function in the context of either divisible goods or pure public goods see [9].

individual utility accrued in the past, as per Eq. 2.2, and is thus delayed by an interval τ. Along with expectations about the future, the two parameters, α and $\widehat{f}_c$, the fraction observed as cooperating, determine how individuals expect the level of cooperation to evolve in time.

For simplicity we assume all members of the group share a common rationality in their method of forming expectations. Specifically, all members expect the game to be of finite duration H, the horizon length[3]. Thus, future returns expected at a time t' from the present are discounted at a rate $e^{-t'/H}$ with respect to immediate expected returns. Secondly, each member expects that their choice of action, when reflected in the net benefits received by the others, will influence future levels of cooperation. Since, however, the decision of one individual affects others' returns by an increment or decrement of only b/n, each member perceives its influence as decreasing with increasing group size. Furthermore, individual changes in strategy are believed to be most effective in encouraging similar behavior when levels of cooperation are high. We postulate that these two effects compound so that each member expects its decision to cooperate or defect to encourage an overall growth or decay in the level of cooperation at a rate proportional to the ratio $\widehat{f}_c/n$. Roughly, then, a member expects its cooperative (defecting) action to stimulate an additional $\widehat{f}_c/n$ members to cooperate (defect) during each subsequent time period.[4]

A mathematical formulation of the manner in which cooperation encourages cooperation and defection encourages defection with nice asymptotic properties now follows. Let $\Delta\widehat{f}_c(t+t')$ denote the expected future difference (at time $t+t'$) between the fraction of agents cooperating and the fraction of those defecting. Member i's choice itself causes an instantaneous difference at $t'=0$ of $\Delta\widehat{f}_c(t, t'=0) = 1/n$. To reflect member i's expectation that during subsequent time steps (of average duration $1/\alpha$), its action will encourage $\widehat{f}_c/n$ additional members per time step to behave likewise, we stipulate that the difference $\Delta\widehat{f}_c(t+t')$ approach 1 asymptotically from its initial value $1/n$ at some given rate. For the purposes of simplicity we will set this rate equal to $e^{-\alpha \widehat{f}_c(t-\tau)t'/n}$, which corresponds to the following deviation:

$$\Delta\widehat{f}_c(t+t') = 1-(1-1/n)\exp\left(-\frac{\alpha\widehat{f}_c(t-\tau)}{n}t'\right). \tag{2.4}$$

We should point out that variations on the precise functional form of the expected deviation $\Delta\widehat{f}_c(t+t')$ would simply cause to the deviation to grow faster or slower with

[3] The concept of a horizon is formally related to a discount δ, which reflects the perceived probability that the game will continue through the next time step. The two are connected through the relation $\sum_{i=0}^{\infty} \delta^i\,() \rightarrow \int_0^{\infty} dt'\, e^{-t'/H}\,()$ which implies $H = \frac{1}{1-\delta}$.

[4] Assigning a functional form to individual expectations, although somewhat arbitrary, is necessary to study dynamics. Variations on the same theme, however, can be seen to yield similar dynamics, provided that the rate at which similar behavior is expected to encourage similar behavior rises monotonically in the ratio $\widehat{f}_c/n$.

increasing t', without yielding significant qualitative changes in the types of dynamical behavior characterizing the interaction.

In summary, member i reevaluates its decision whether or not to contribute to the production of the good at an average rate α, using its knowledge of $\widehat{f}_c(t-\tau)$ and following its expectations about the future. Using its prediction of how it expects f_c to evolve in relation to its choice and discounting the future appropriately, member i then makes its decision on whether to cooperate or defect.

Putting it all together, member i perceives the advantage of cooperating over defecting at time t to be the net benefit

$$\Delta\widehat{B}_i(t) = H(b-c) - \frac{Hb(n-1)}{n + H\alpha\widehat{f}_c(t-\tau)}. \tag{2.5}$$

Member i cooperates when $\Delta\widehat{B}_i(t) > 0$, defects when $\Delta\widehat{B}_i(t) < 0$, and chooses at random between defection and cooperation when $\Delta\widehat{B}_i(t) = 0$, basing its decision on the fraction of the group it perceives to have cooperated at a time τ in the past, $\widehat{f}_c(t-\tau)$. These criteria reduce to the following condition for cooperation at time t:

$$f_{crit} \equiv \frac{1}{H\alpha}\left(\frac{nc-b}{b-c}\right) < \widehat{f}_c(t-\tau). \tag{2.6}$$

Since $\widehat{f}_c(t-\tau)$ is a mixture of two binomially distributed variables, Eq. 2.6 provides a full prescription of the stochastic evolution for the interaction. In particular, member i cooperates with probability $P_c(f_c(t-\tau))$ that it perceives cooperation as maximizing its expected future accumulated utility, given the actual attempted level of cooperation $f_c(t-\tau)$.

Thus, the individuals engage in conditional cooperation, cooperating when the fraction perceived as cooperating is greater than the critical amount, f_{crit} and defecting when the perceived fraction cooperating is less than f_{crit}. Because of the nature of the individuals' expectations, this behavior, reminiscent of a generalized tit-for-tat, emerges from individually rational choices. In this way, conditional cooperation is a rational strategy. Furthermore, conditional cooperation is a more general type of strategy than might be expected from the explicit specification of the individuals' beliefs. As a result, it holds for a wider range of models than considered here, once one allows for variation in the functional form of f_{crit}.

Group behavior

A deterministic continuous version of the stochastic discrete interaction specified above can be obtained assuming: (1) the size, n, of the group is large; and (2) the average value of a function of some variable is well approximated by the value of the function at the average of that variable. The model developed below will be useful in discovering the Nash equilibria and determining their stability characteristics.

From this point on, we specialize to the symmetric case $p = q$, in which an individual is equally likely to effectively defect when intending to cooperate as to cooperate when intending to defect. (Later we will comment on the effect of this restriction on the generalizability of our results.) By the Central Limit Theorem, for large n, the random variable $\widehat{f}_c$ tends in law to a Gaussian distribution with mean $\langle \widehat{f}_c \rangle = pf_c + (1-p)(1-f_c)$ and variance $\sigma^2 = p(1-p)/n$. Under our assumptions, the mean probability that $\widehat{f}_c > f_{crit}$ thus becomes

$$\rho_c(f_c) = \frac{1}{2}\left\{1 + \text{erf}\left[\frac{\langle \widehat{f}_c \rangle - f_{crit}}{\sqrt{2}\sigma}\right]\right\}. \tag{2.7}$$

The evolution of the number of agents cooperating in time is then described by the dynamical equation [16]

$$\frac{df_c}{dt} = -\alpha[f_c(t) - \rho_c(f_c(t-\tau))], \tag{2.8}$$

where α is the reevaluation rate and τ is the delay parameter, as defined earlier.

The equilibrium points f_c^0 of the interaction described by the above equation are obtained by setting the right hand side to zero[5]. In this case they are given by the solutions to

$$\rho(f_c^0) = f_c^0. \tag{2.9}$$

Solving the above equation yields the critical sizes beyond which cooperation can no longer be sustained. In the case of perfect certainty (p=q=1), these critical sizes can be expressed in simple analytical form. Thus, a group will no longer sustain global cooperation if it exceeds a value n^* given by

$$n^* = H\alpha\left(\frac{b}{c} - 1\right) + \frac{b}{c}. \tag{2.10}$$

[5]Furthermore, linear stability analysis of Eq. 2.8 shows that the stability of the equilibrium points is independent of the value of the delay τ, and of the reevaluation rate α. Thus, the asymptotic behavior of the group interaction does not depend on the delay. Moreover, the equilibrium points belong to one of two types: stable fixed point attractors or unstable fixed point repellors, due to the linearity of the condition for cooperation (Eq. 2.6).

Similarly, cooperation is the only possible global outcome if the group size falls below a second critical size n_{min}:

$$n_{min} = \frac{b}{2c} + \frac{1}{2c}\sqrt{b^2 + 4H\alpha c(b-c)}. \tag{2.11}$$

Notice that these two critical sizes are not equal; in other words, there is a range of sizes between n_{min} and n^* for which either cooperation or defection is a possible outcome, depending on the initial conditions.

An estimate of the possible sizes can be obtained, for example, if one assumes a horizon H=50 (which corresponds to a termination probability δ=0.98), α=1, b=2.5 and c=1. In this case one obtains $n^* = 77$ and $n_{min} = 10$, a significant group size. Observe that an increase in the horizon length would lead to corresponding increases in the critical sizes.

3 The Critical Mass

The Ω function formalism

The model studied in the previous section dealt with the average properties of a collection of agents having to choose between cooperation and defection. Since the asymptotic behavior generated by the dynamics was in the form of Nash equilibria or fixed points, it is of interest to ask about the evolution of fluctuations away from equilibrium state in the presence of uncertainty. These fluctuations are important for two reasons: (1) the time necessary for the system to relax back to equilibrium after small departures in the number of agents cooperating or defecting might be long compared to the time-scale of the collective task to be performed, or the measuring time of an outside observer; (2) large enough fluctuations in the number of defecting or collaborating agents can shift the state of the system from cooperating to defecting and *vice-versa*. If that is the case, it becomes important to know how probable these large fluctuations are, and how they evolve in time.

In what follows we will use a formalism introduced by Ceccatto and Huberman [12] that is well suited for studying fluctuations away from the equilibrium behavior of the system. This formalism relies on the existence of an optimality function, Ω, that can be constructed from knowledge of the density dependent utilities. This function has the important property that its local minima give the Nash equilibria of the system as the most probable configurations of the system. Depending on the complexity of the function, several Nash equilibria can exist, with the overall global minimum producing the optimal state of the system.

Specifically, the equilibrium probability distribution $P_e(f_c)$ is given by

$$P_e = C \exp\left[-n\Omega(f_c)\right], \tag{3.1}$$

where the optimality function Ω for our model of ongoing collective action is given by

$$\Omega(f_c) = \int_0^{f_c} df'_c \left[f'_c - \rho_c(f'_c)\right] \tag{3.2}$$

in terms of the mean probability $\rho_c(f_c)$ that cooperation is preferred. Thus, the optimal configuration corresponds to the value of f_c at which Ω reaches its global minimum.

Within this formalism it is easy to study the dynamics of fluctuations away from these minima. First, consider the case where there is a single Nash equilibrium (which can be either cooperative or defecting). As was shown, fluctuations away from this state relax back exponentially fast to the equilibrium point, with a characteristic time of the order of $1/\alpha$, which is the average evaluation time for the individuals. Second, and more interestingly, is the situation when there are multiple Nash equilibria, with the global minimum of the Ω function denoting the optimal state of the system. This is illustrated schematically in Fig. 1.

If the system is initially in a Nash equilibrium which corresponds to the global minimum (*e.g.*, state *A*), fluctuations away from this state will relax back exponentially

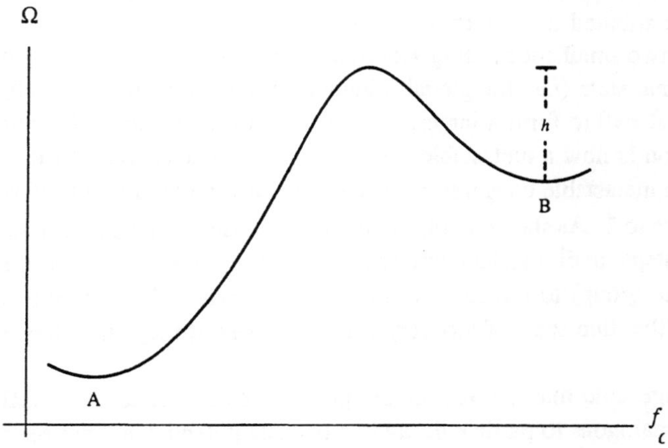

Fig. 1. Schematic sketch of the optimality function Ω *vs.* f_c, the fraction of agents cooperation. The global minimum is at *A*, local minimum at *B*. *h* is the barrier height separating state *B* from *A*.

fast to that state. But if the system is initially trapped in a metastable state (state B), *i.e.*, a minimum which is not the global one, the dynamics away from this state is both more complicated and interesting. As was shown by Ceccatto and Huberman, whereas for short times fluctuations away from the local minimum relax back to it, for longer times a giant fluctuation can take place, whereby a large fraction of the agents switching strategies can push the system over the barrier maximum. Once this critical mass is reached, the remaining agents rapidly switch into the new strategy that corresponds to the optimal Nash equilibrium and the system slides into the optimal state.

The time scales over which this whole process takes place is also of interest, for the time to nucleate a giant fluctuation is exponential in the number of agents. However, when such transitions take place, they do so very rapidly — the total time it takes for all agents to do the crossing is logarithmic in the number of agents. Since the logarithm of a large number is very small when compared to an exponential of the same number, the theory predicts that nothing much happens for long times, but when it does, it happens very fast.

The process of escaping from the metastable state depends on the amount of imperfect knowledge that individuals have about the state of the system, in other words, on what other agents are doing. In the absence of imperfect knowledge the system would always stay in the local minimum downhill from the initial conditions, since small excursions away from it by a few agents would reduce their utility. It is only in the case of imperfect knowledge that many individuals can change their behavior. This is because, in evaluating the number of members cooperating, imperfect knowledge amounts to occasional large errors in the individual's estimation of the actual number cooperating.

Results of Monte Carlo simulations, which were conducted in asynchronous fashion, confirm these theoretical predictions. Each individual decides to cooperate or defect based on the criterion given in Eq. 2.6. Uncertainty enters since these decisions are based on perceived levels of cooperation which differ from the actual attempted amount of cooperation in a way distributed as a mixture of binomials.

For example, consider two small cooperating groups of size n=6, with horizon length H=9.5, for which the optimal state (*i.e.*, the global minimum of the optimality function) is cooperation, that merge at t=0 to form a larger, cooperating group of size n=12. For the larger group, cooperation is now a metastable state: no one individual will find it to its benefit to defect and the metastable cooperative state can be maintained for very long times, especially if p is close to 1. As shown in Fig. 2, in this one case mutual cooperation lasts for about 4000 time steps, until a sudden transition (of duration proportional to the logarithm of the size of the group) to mutual defection occurs, from which the system will almost never recover (the time scale of recovery is many orders of magnitude larger than the crossover time).

Determining the average time that it takes for the group to crossover to the global minimum is a calculation analogous to particle decay in a bistable potential and has been performed many times [17]. The time, t, that it takes for a group of size n to cross over from a metastable Nash equilibrium to the optimal one is given by

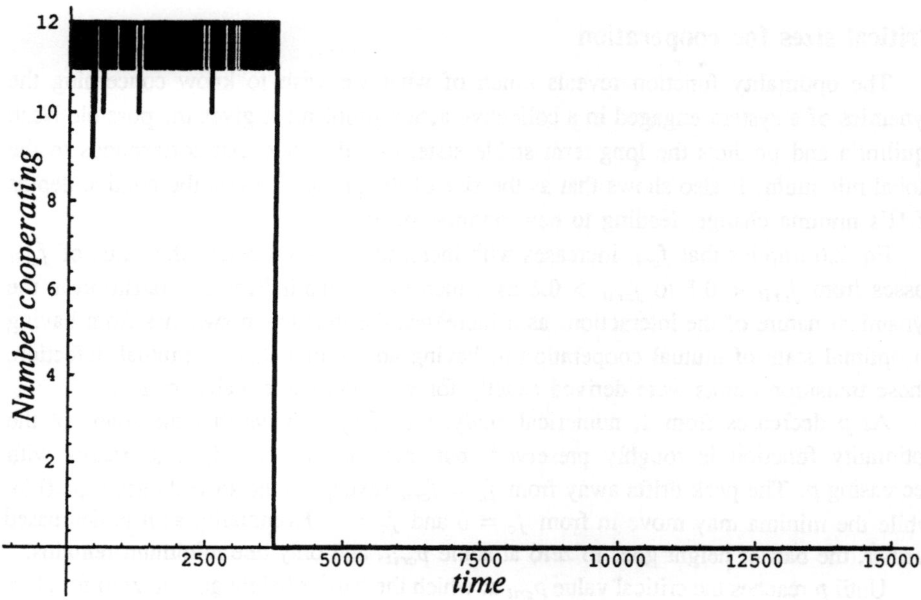

Fig. 2. At t=0, two cooperating groups of size n=6 merge to form a larger, cooperating group of size n=12. All agents have horizon length H=9.5, with p=0.93, b=2.5, c=1, α=1, and τ=1. For these parameters, cooperation is the optimal state for a group of size n=6, but for the combined group of size n=12, cooperation is metastable. Indeed, as the figure shows, metastable cooperation persists for almost 4,000 time steps in this example. The average crossover time is about 5,000 time steps, ranging from less than 1,000 to over 10,000 time steps. Uncertainty (p less than one) ensures that eventually a large fluctuation in the perceived number of agents cooperating eventually takes the group over into a state of mutual defection, which is optimal.

$$t = \text{constant } e^{nh/\sigma}, \tag{3.3}$$

with h the height of the barrier as shown in Fig. 1 and σ a measure of the imperfectness of the individuals' knowledge. We should point out, however, that in our model the barrier height itself also depends on n, H, and p, making simple analytical estimates of the crossover time considerably more difficult.

Further simulations of the example given above show that the average crossover time in that case is about 5,000 time steps, although it can range from less than 1,000 to over 10,000 time steps. The exponential dependence of the crossover time on the amount of uncertainty can be seen by running the same system, but with different amounts of error. In the example above, p equals 0.93. However, if the amount of error increases so that p now equals 0.91, say (thus reducing the height of barrier between cooperation and defection by 21%), the crossover to defection typically occurs within hundreds of time steps, instead of thousands.

Critical sizes for cooperation

The optimality function reveals much of what we wish to know concerning the dynamics of a system engaged in a collective action problem: it gives the possible Nash equilibria and predicts the long-term stable state, *i.e.*, the state that corresponds to the global minimum. It also shows that as the size of the group changes, the relative depths of Ω's minima change, leading to new optimal states.

Eq. 2.6 implies that f_{crit} increases with increasing n. As a result, the value of f_{crit} passes from $f_{crit} < 0.5$ to $f_{crit} > 0.5$ as n increases. This indicates a transition in the dynamical nature of the interaction: as n increases, the interaction switches from having an optimal state of mutual cooperation to having an optimal state of mutual defection. These transition points were derived exactly for $p=1$ and numerically for $p<1$.

As p decreases from 1, numerical analysis of $\Omega(f_c)$ shows that the shape of the optimality function is roughly preserved, but that the barrier height decreases with decreasing p. The peak drifts away from $f_c = f_{crit}$ (except in the special case $f_{crit}=0.5$), while the minima may move in from $f_c = 0$ and $f_c = 1$. Eventually, as p is decreased further, the barrier height goes to zero at some p_{crit}, and only one minimum remains.

Until p reaches the critical value p_{crit} at which the barrier height goes to zero for all n, the nature of the system's equilibrium points remain similar to the $p=1$ case. Specifically, for $p < p_{crit}$, three critical values can be obtained: (1) $n_{min}(p)$, the minimum group size below which cooperation is the only fixed point; (2) $\tilde{n}^*(p)$, the critical size below which cooperation is the optimal state; and (3) $n^*(p)$, an upper bound above which cooperation is not sustainable. The values $n_{min}(p)$, $\tilde{n}^*(p)$, and $n^*(p)$ can be determined numerically, demonstrating the emergence of four levels in group size corresponding to very different resolutions of the collective action problem:

$n \leq n_{min}(p)$	one equilibrium point – mostly cooperative;
$n_{min}(p) < n \leq \tilde{n}^*(p)$	mostly cooperative optimal state, mostly defecting metastable state;
$\tilde{n}^*(p) < n < n^*(p)$	mostly defecting optimal state, mostly cooperative metastable state;
$n \geq n^*(p)$	one equilibrium point – mostly defecting.

Analysis of the optimality function yields the values $n_{min}(p)$, $\tilde{n}^*(p)$, and $n^*(p)$, along with p_{crit}. These values were found for the case $H=50$ (which corresponds to discount rate $\delta = 1-\frac{1}{H} = 0.98$), $\alpha=1$, $\tau=1$, benefit to group of individual cooperation $b=2.5$, and personal cost of cooperation $c=1$. The resulting phase diagram delineating the regions of different resolutions to the conflict is shown in Fig. 3. At $p=1$, $n^* = 77, \tilde{n}^* = 40$, and $n_{min} = 10$. When $p > p_{crit} = 0.59$, the uncertainty is high enough that all structure in

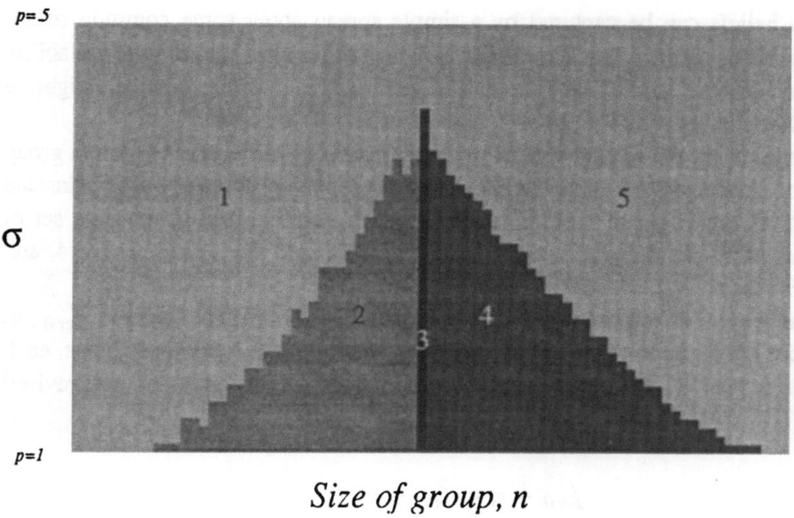

Fig. 3. Diagram delineates regions corresponding to different resolutions of collective action problem for parameter values H=50 and α=1. The amount of error, σ, increases vertically, and the size of the group increases horizontally. In region 1, $n \leq n_{min}(p)$, there is one equilibrium point – mostly cooperative; in region 2, $n_{min}(p) < n < \tilde{n}^*(p)$, mostly cooperative is the optimal state, while mostly defecting is a metastable state; in region 3, $n = \tilde{n}^*(p)$,the system is bistable with mostly cooperative and mostly defecting both optimal states; in region 4, $\tilde{n}^*(p) < n < n^*(p)$, mostly defecting is the optimal state, while mostly cooperative is a metastable state; and in region 5, $n^*(p) \leq n$, there is again only one equilibrium point – mostly defecting. There is no sharp boundary between regions 1 and 5 for high levels of uncertainty. Note that region 3 actually has zero width.

the optimality function is washed out and only one equilibrium point exists for group of all sizes. For groups of size less than 40 operating within such a high level of uncertainty, the interaction evolves to a mixed equilibrium that is more cooperative than defective. The balance reverses itself when the group size exceeds 40.

The diagram in Fig. 3 shows that for the symmetric case $p = q$, $n^*(p)$ is a decreasing function of p while $n_{min}(p)$ is an increasing function of p and $\tilde{n}^*(p)$ is a constant. In general $(q \neq p)$, $n^*(p)$ remains a decreasing function; however, the functional forms of $\tilde{n}^*(p)$ and $n_{min}(p)$ depend on q. For example, q=0 yields decreasing functions $\tilde{n}^*(p)$ and $n_{min}(p)$.

4 Diversity and Cooperation

In the first pass at studying collective action problems, we treated all individuals as identical. We now drop that assumption and consider how diversity enters and how behavior changes as a result. As we discuss below, two qualitatively different forms of diversity must be studied. The first type reflects diversity in groups of agents whose differences in beliefs can be captured by a simple spread about some common belief. Thus, the individuals on the whole are similar, but have differences that capture variability in preferences and other additional factors. For example, each group member might be said to have horizon length $H = 10 \pm 2$, instead of $H=10$ exactly.

On the other hand, the second type of diversity represents differences within a group that cannot be accounted for by a simple variance about an average value. Instead the group acts as the union of several subgroups each characterized by its own set of beliefs. In this case, one subgroup might have horizon length $H=2$, another $H=4$, and yet another $H=6$.

A general way of incorporating diversity is to allow the critical fraction, f_{crit}, in the criterion for cooperation (Eq. 2.6) to vary from individual to individual. Thus, each member i has a bias b_i, and decides whether or not cooperate based on the revised condition

$$f_{crit} + b_i < \widehat{f}_c(t - \tau). \tag{4.1}$$

The notation we use to denote the biased value of the critical fraction will be $f_{crit}^{b_i} \equiv f_{crit} + b_i$.

Diversity as a form of uncertainty

We examine first the case in which the group's diversity can be modeled by a Gaussian distribution. If we take the biases $\{b_i\}$ to be distributed normally with mean zero, then the critical size beyond which cooperation cannot be maintained remains the same as without diversity. Distributions with non-zero mean would alter the nature of the equilibrium points; so zero mean distributions offer the best means to isolate the effect of adding diversity.

In addition, the distribution of biases $\{b_i\}$ can be used to represent a diversity in the individuals' horizons or in their benefit-cost ratios b/c as long as $\sigma \ll f_{crit}$.. Thus, describing diversity by adding a bias to the criterion for cooperation of Eq. 2.6 is a more general description of diversity than might appear at first glance.

In order to understand the qualitative effect of introducing diversity in this manner, we write the mean probability $\rho_c^i(f_c)$ that member i will choose to cooperate as

$$\rho_c^i(f_c) = \frac{1}{2}\left\{1 + erf\left[\frac{\langle\widehat{f}_c\rangle - (f_{crit} + b_i)}{\sqrt{2}\sigma}\right]\right\}, \tag{4.2}$$

where $\langle\widehat{f_c}\rangle = pf_c + (1-p)(1-f_c)$ and $\sigma = \sqrt{p(1-p)/n}$. The dynamical equation describing the evolution of the system then becomes

$$\frac{df_c}{dt} = -\alpha\left[f_c(t) - \frac{1}{n}\sum_{i=1}^{n}\rho_c^i(f_c(t-\tau))\right]. \tag{4.3}$$

Letting

$$\tilde{\rho}_c(f_c) = \frac{1}{n}\sum_{i=1}^{n}\rho_c^i(f_c(t-\tau)). \tag{4.4}$$

and linearizing about $\langle\widehat{f_c}\rangle = f_{crit}$, we obtain

$$\tilde{\rho}_c(f_c) = \frac{1}{2}\left\{1 + erf\left[\frac{\langle\widehat{f_c}\rangle - f_{crit}}{\sqrt{2}\tilde{\sigma}}\right]\right\}, \tag{4.5}$$

with the renormalized value of uncertainty $\tilde{\sigma}$ given by

$$\tilde{\sigma} \rightarrow \sqrt{\sigma^2 + \sigma'^2}. \tag{4.6}$$

This approximation assumes that the $\{b_i\}$ are distributed normally with mean zero and standard deviation σ'.

Thus, taking this first form of diversity into account simply renormalizes the amount of noise in the system as parametrized by σ in the denominator of the error function. Note, however, that imperfect information also enters in the calculation of the expected value of f_c: $\langle\widehat{f_c}\rangle = pf_c + (1-p)(1-f_c)$. Consequently, adding diversity is not identical to decreasing p away from 1.

From this analysis, it appears that diversity among the agents will shorten the lifetime of the metastable states described in the previous section, while the transitions remain abrupt. Computer experiments verify this prediction, with a sample simulation given in Fig. 4. Diversity is modeled as a spread about f_{crit} among the agents; thus, agents differ from one another in their likelihood of cooperating *vs.* defecting. In this example, the individual biases to f_{crit} are distributed normally with standard deviation set equal to the amount due to imperfect information $\left(\sigma' \equiv \sigma = \sqrt{p(1-p)/n}\right)$. As a result of the diversity among the individuals in the group, the average crossover time becomes about 1,300 time steps, ranging from fewer than 100 time steps to over 2,000. Without

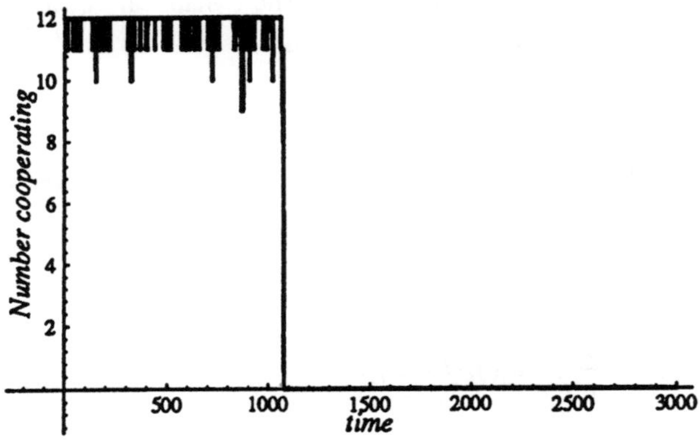

Fig. 4. Group of size n=12, with H=9.5, p=0.93, b=2.5, c=1, α=1, and τ=1 with diversity. Diversity is modeled by adding individual biases to the value of $f_{crit} = 0.6667$. These biases are distributed normally with mean zero and standard deviation $\sigma' \equiv \sigma = \sqrt{p(1-p)/n}$ (the "noise" due to diversity is set equal to the noise due to uncertainty). The moderate amount of diversity is responsible for a transition rate to defection about four times as fast as that for a uniform group (see, for example, Fig. 2 — note the difference in time scales).

diversity, the average crossover time is about 5,000 time steps (see, for example, Fig. 2 — note in particular the difference in time scales), ranging from less than 1,000 time steps to over 10,000.

Stages to cooperation

We next consider the second form of diversity, representing large differences in beliefs between various subgroups within a group. This type of diversity follows a multimodal distribution and does not lend itself to the renormalization of uncertainty performed above. Instead, the summed error function in Eq. 4.3 exhibits a sequence of steps. For example, if in a group of size 12 there is one subgroup biased towards cooperation with $f^{b_1}_{crit} = 0.1333$, another with $f^{b_2}_{crit} = 0.4667$ and yet another biased towards defection with $f^{b_3}_{crit} = 0.8$, then there are three steps in the summed error function of Eq. 4.4. Each step represents the likelihood that an individual from each respective subgroup will find it worthwhile to cooperate. There can now be multiple local minima in which some of the subgroups sustain cooperation while the others do not. As always, the global minimum is the long-term preferred state of the system.

In the example given in the preceding paragraph, overall cooperation can be achieved in step-wise fashion. If the group begins in an uncooperative state, the subgroup with $f^{b_1}_{crit} = 0.1333$ is the first to make the transition to cooperation, followed by the subgroup with $f^{b_2}_{crit} = 0.4667$ and finally by the subgroup with $f^{b_3}_{crit} = 0.8$. These steps can be seen in Fig. 5.

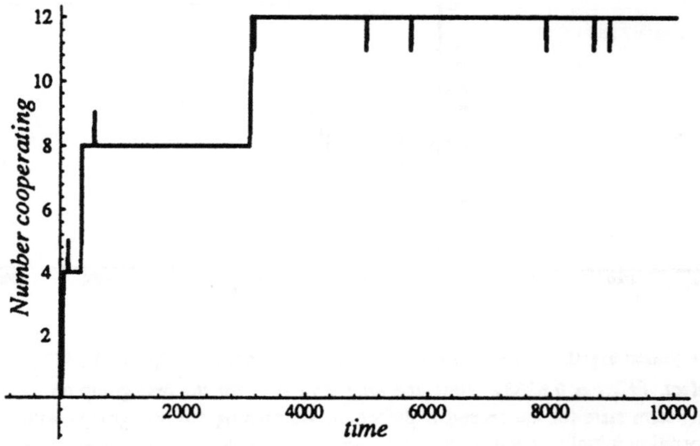

Fig. 5. *Stages towards cooperation.* A group of size 12 consists of three subgroups. The first is greatly biased towards cooperation with $f^{b_1}_{crit} = 0.1333$. The second subgroup has $f^{b_2}_{crit} = 0.4667$, while the third is biased towards defection with $f^{b_3}_{crit} = 0.8$. The figure shows the stepwise transition to mutual cooperation: each subgroup makes the transition to cooperation in turn. Parameter values are p=0.98, b=2.5, c=1, n=12, and H=12.

Compare this example to the uniform case in which all individuals have $f_{crit} = 0.4667$ (the average critical fraction for the diverse group above). This system is now almost bistable, with both cooperation and defection about equally likely to be the long-term stable state, although cooperation is more likely. In this scenario, then, diversity greatly increases the likelihood of attaining a cooperative state. Other scenarios can be envisioned, however, which accomplish the reverse, making defection more likely with diversity than without.

Another interesting consequence of the effect of multimodal diversity is the following. Imagine combining two groups, each composed of 6 individuals. The first group has beliefs such that all members have $f^{b}_{crit} = 0.3667$. Thus, the long-term stable state for this group is one of cooperation. The second group, on the other hand, is unbiased with $f_{crit} = 0.6667$, and thus tends towards mutual defection. Putting these groups together, then, we have an initial state with 6 of the 12 individuals cooperating. This turns out to be a metastable state of the combined system. Eventually, however, the group undergoes a transition to either complete cooperation or complete defection. Both are equally likely — the two possibilities are illustrated in Fig. 6. Thus, either the defecting agents are persuaded to cooperate (although they generally wouldn't when in a group to themselves even though the size of this subgroup is smaller); or the originally cooperating agents lose their incentive to do so and defect.

This example invites comparison to the situation in which all members have $f_{crit} = 0.5666$. Such a group would be nearly bistable, although with a long-term tendency towards defection. A group with initially half its members cooperating would most of time evolve rapidly to a state of mutual defection. With diversity as above, on the other

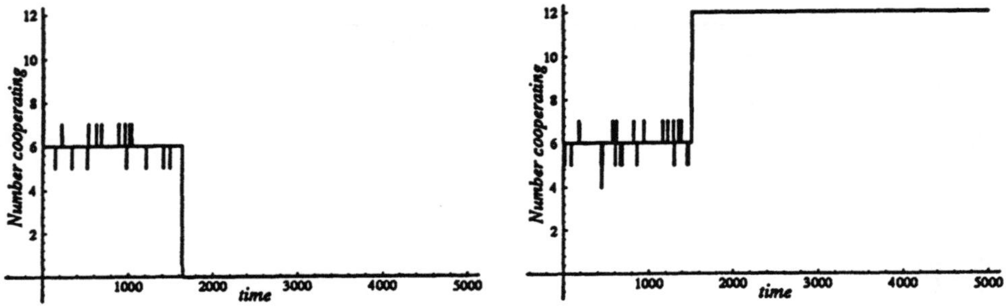

Fig. 6. Two groups of size 6 are joined together, one with a bias towards cooperation $\left(f_{crit}^{b_1} = 0.3667\right)$, the second with a tendency to defect. $(f_{crit} = 0.6333)$. Thus, the long-term state for the first group on its own is cooperation, while the long-term state for the second is defection. Combining the two groups with the six agents biases towards cooperation initially cooperating and the second six defecting, the figures show that overall behavior can go either way. The initial state is itself metastable, but fluctuations insure that eventually the group will go over to either mutual cooperation or mutual defection, with both possibilities equally likely. Either the defecting agents are swept over into cooperation or the cooperative agents collapse into defection. Parameter values are p=0.99, b=2.5, c=1, n=12, H=12.

hand, the two stable states of defection and cooperation become separated by a metastable state in which half of the group cooperates, and the group can now remain stuck in a situation in which half of the group free rides on the other half. For this particular choice of diversity, transitions to either overall cooperation or defection are equally likely. Thus, the diversity in the group has improved the group's chances of mutual cooperation over those of a uniform group with the same average f_{crit}.

5 Discussion

In this paper we presented and studied a model of ongoing collective action among intentional agents which make choices that depend not only to the past but also on their expectations. This model maps collective action problems onto an iterated n-person prisoners' dilemma in an uncertain world. Individuals interact dynamically, making decisions based both on individual preferences and on their expectations as to how their choices will affect other agents in the future. This is a significant departure from previous studies of collective action problems in which the finiteness of the interaction is the only shadow cast by the future.

Using a dynamical formulation of these interactions we derived the magnitudes of group sizes that are capable of supporting cooperation and established the existence of regimes with multiple Nash equilibria for groups of identical agents. Extensive computer experiments also allowed us to confirm the analytically derived behavioral diagrams whose distinct regions correspond to different resolutions of the collective action problem.

We also studied stochastic fluctuations in the number of individuals cooperating or defecting using a thermodynamic-like formalism. This allowed us to discover the existence of very long transient states in which collaboration can persist in groups whose sizes are too big to support it indefinitely. Moreover, the onset of overall cooperation takes place in a sudden and unexpected way. Likewise, defection can appear out of nowhere in very large, previously cooperating, groups. These outbreaks, which were also confirmed by computer experiments, mark the end of long transient states in which defection or cooperation persists in groups that cannot sustain it indefinitely. Since these discontinuities are not overly sensitive to the analytical form of the expectations, we expect that they might be observed in more complex social settings.

Relaxing the assumption that the individuals were identical in their beliefs and preferences, we then elucidated the effect of diversity on the dynamics of collective action. One type of diversity, represented as a spread about some common average belief among individuals, was shown to act as a source of extra uncertainty, thus shortening the time to an outbreak without affecting its abrupt nature. A second form of diversity, which occurs when a group consists of several subgroups each with its own distinct beliefs was also studied. In this case, we showed that cooperation (or defection) can be achieved by the group in demarcated stages. In general, an additional metastable state appears for each new subdivision within the group.

The implications of this work for the study of cooperation in organizations can be briefly stated. In order to achieve spontaneous cooperation on a global scale, an organization should be structured into small subunits made up of individuals with a well-designed diversity of beliefs. The small size of the units allows for the emergence of sustained cooperation: diversity hastens its appearance. To the extent that these two elements, smallness and diversity, can be preserved in a hierarchical structure, we anticipate that sustained cooperation can be achieved in more complex systems as well.

References

[1] Mancur Olson. *The Logic of Collective Action*. Harvard University Press, Cambridge, 1965.

[2] Garrett Hardin. The tragedy of the commons. *Science*, 162:1243–1248, 1968.

[3] Michael Taylor. *Anarchy and Cooperation*. John Wiley and Sons, New York, 1976.

[4] Thomas C. Schelling. *Micromotives and Macrobehavior*. W. W. Norton and Company, Inc., United States, 1978.

[5] Russell Hardin. *Collective Action*. Johns Hopkins University Press, Baltimore, 1982.

[6] Michael Taylor. *The Possibility of Cooperation*. Cambridge University Press, Cambridge, 1987.

[7] David D. Friedman. *Price Theory*. South-Western Publishing Co., Ohio, 1990.

[8] Pamela E. Oliver and Gerald Marwell. The paradox of group size in collective action: A theory of the critical mass. ii. *American Sociological Review*, 53:1–8, 1988.

[9] Jonathan Bendor and Dilip Mookherjee. Institutional structure and the logic of ongoing collective action. *American Political Science Review*, 81(1):129–154, March 1987.

[10] Russell Hardin. Collective action as an agreeable n-prisoners' dilemma. *Behavioral Science*, 16(5):472–481, 1971.

[11] Robert Axelrod and William D. Hamilton. The evolution of cooperation. *Science*, 211:1390–1396, 1981.

[12] H. A. Ceccatto and B. A. Huberman. Persistence of nonoptimal strategies. *Proc. Natl. Acad. Sci. USA*, 86:3443–3446, 1989.

[13] Natalie S. Glance and Bernardo A. Huberman. The outbreak of cooperation. *Journal of Mathematical Sociology*, 17, 1992.

[14] Reinhardt Selten. Re-examination of the perfectness concept for equilibirium points in extensive games. *International Journal of Game Theory*, 4:25–55, 1975.

[15] Herbert Simon. *The Sciences of the Artificial.* Massachusetts Institute of Technology, Cambridge, 1969.

[16] Bernardo A. Huberman and Tad Hogg. The behavior of computational ecologies. In B. A. Huberman, editor, *The Ecology of Computation*, pages 77–115. North-Holland, Amsterdam, 1988.

[17] Masuo Suzuki. Scaling theory of transient phenomena near the instability point. *Journal of Statistical Physics*, 16:11–32, 1977.

On the Application of Synergetics to Social Systems

W. Wischert, A. Wunderlin

Institut für Theoretische Physik und Synergetik,
Universität Stuttgart, Pfaffenwaldring 57/4,
D-7000 Stuttgart 80, Fed. Rep. of Germany

Abstract: On a purely macroscopic level we propose a method to model social systems near instabilities. Our suggestion is based on fundamental principles and results of synergetics. Two different examples are discussed in detail.

1 Introduction

It is our aim to propose a general procedure for applying synergetics to social systems. The method is justified from the fundamental result of synergetics that there are new laws governing, in their own right, the macroscopic behavior of complex systems. These laws turn out to be universal in the vicinity of critical regions of a complex system [1, 2]. They allow us to give a qualitative mathematical description in terms of order parameters and to make predictions about possible evolving dynamical behavior on macroscopic scales. Our approach is phenomenological in the sense that we have to develop identification strategies for the variables and to give an interpretation to the emerging unstable behavior. The quality of the models strongly depends on the identification strategy. In this way a subjectivistic element is introduced into our reasoning. The results, however, can be analyzed on the level of empirical knowledge which is available in social sciences. The models we shall consider here are purely deterministic in nature. However, a generalization including stochastic behavior appears to be straightforward.

We would like to emphasize that our method is located purely on a macroscopic level. It adds a further way of arguing to the quite different approaches which are based on a microscopic or mesoscopic treatment [3–6]. In these cases many variables, time lags, and interaction mechanisms have to be introduced and the interpretation of the results may accordingly become extremely difficult. It even appears that, because of numerical difficulties, many structures inherent in the original equations might be lost in a simulation. The order parameter concept of synergetics provides us with a tool which has a quite different quality. In the vicinity of an instability only few order parameters are needed and the interpretation of the instability becomes much simpler. Furthermore, it is important to note that the order parameter concept still allows for a nonlinear treatment of the instabilities.

Our article is organized as follows: In Sect. 2 we summarize important results of synergetics which provide the basis for our subsequent consideration. In Sect. 3 we briefly describe the method of phenomenological synergetics and discuss identification strategies. Sect. 4 then tries to develop a model of a society by introducing the concept

Springer Series in Synergetics, Vol. 62 **Interdisciplinary Approaches to Nonlinear Complex Systems** – Eds.: H. Haken and A. Mikhailov

of collective modes. As special examples we then treat Weidlich's model of opinion formation [4–6] from our point of view. We then develop another model which has various applications in quite different fields like sociology, economy, ecology, etc. In our case we choose the behavior of a company. In Sect. 5 we draw important conclusions and, in particular, discuss the value of predictions from a nonlinear point of view.

2 Some Basic Results of Synergetics

Here it is our concern to summarize some fundamental results of synergetics which are needed for later considerations and keep the article self-contained. The aim of synergetics is to describe processes of spontaneous self-organization and cooperation in complex systems. These systems are built from many subsystems which themselves can be complicated objects. We especially meet this situation when we try to apply the methods of synergetics to social systems where, for example, the individual human and/or groups of humans are considered as subsystems. General properties of the subsystems are their nonlinear dynamics as well as their nonlinear interactions. Furthermore, the systems of synergetics are regarded as open systems. The influence from outside is measured by a certain set of control parameters $\{\sigma_i\}$. Processes of self-organization are observed as spatial, temporal, or spatio-temporal macroscopic patterns. They are described by the order parameters. Another result of self-organization on macroscopic scales can be realized in a special functioning of a complex system.

The interdisciplinary approach of synergetic theory can be justified from the general observation that the behavior of such systems on macroscopic scales is independent of any details of the microscopic nature of the subsystems and their interactions. Indeed, the macroscopic action is governed by the order parameters and their dynamics which can be classified from purely macroscopic reasoning [1, 2]. In order to clarify these statements we briefly describe the important steps, starting from the microscopic level, in deriving the macroscopic order parameter equations.

A complex system is described by a state vector q and the components of q are the complete set of variables which characterize the state of each individual subsystem. The state vector evolves in time t. This can be modelled for a large class of systems by an evolution equation of the following type

$$\frac{d}{dt}q(t) = N\left(q(t), \{\sigma_i\}\right) , \tag{1}$$

where N is a nonlinear vector field which reflects the behavior and interactions of the subsystems. The external influences are taken into account by the dependence of N on the control parameters. In more general cases spatial dependences are also allowed and the components of q have to be considered as fields depending on space and time. Furthermore, fluctuations may be added which represent a possibly incomplete knowledge about the states of the subsystem, uncertainties in control, etc. Here, however, we shall confine ourselves to the simplest case which is expressed through (1).

Equation (1) is in general quite complicated and cannot be solved completely in the whole state space Γ spanned by the set of possible vectors $q(t)$. We therefore have to restrict ourselves to local concepts for analyzing the behavior of the system. To this end we shall consider a reference state $q_0(t)$ and its neighborhood. We assume that the reference state has the properties of an attractor and is a comparably low dimensional object in Γ. In order to explore the behavior of our system in the neighborhood of $q_0(t)$ we look for the time development of small deviations from the reference state, that is, we make the ansatz

$$q(t) = q_0(t) + \delta q(t) \tag{2}$$

and consider $\delta q(t)$ as a small quantity. As a result we may linearize the equation of δq in the vicinity of the reference state $q_0(t)$. We obtain

$$\frac{d}{dt}\delta q(t) = L\left(q_0(t), \{\sigma_i\}\right)\delta q(t), \tag{3}$$

where L is a linear matrix independent of δq. L can immediately be derived from the vector field N by standard methods [1, 2]. We now assume that we can construct a complete set of eigenvectors $v^{(j)}(t)$ of L corresponding to (3). These eigenvectors allow us to decompose an arbitrary deviation $\delta q(t)$ into elementary collective deviations along the directions of the eigenvectors

$$\delta q(t) = \sum_{j=1}^{n} \xi_j(t) v^{(j)}(t). \tag{4}$$

In (4) n denotes the dimension of the state space Γ and $\xi_j(t)$ the excitation of the system along the directions prescribed by the $v^{(j)}(t)$. We note that the introduction of the eigenvectors $v^{(j)}$ is of crucial importance. In the realm of synergetics they are considered as the collective modes or patterns of the system. This can be substantiated as follows: Whereas the original equation (1) is formulated on the basis of the variables of the single subsystems, we can now give a new formulation which is based on these collective patterns and describes the dynamical behavior of the system in terms of these different collective patterns. Inserting the ansatz (4) into (1) we obtain an equation for the amplitudes $\xi_j(t)$ which has the following form

$$\frac{d}{dt}\xi_i(t) = \Lambda_{ij}\xi_j(t) + \text{nonlinear terms}. \tag{5}$$

Here we have assumed that the time dependence of the matrix $L(q_0(t))$ is carried by the eigenvectors leaving us with a constant matrix Λ.

We now summarize the results by discussing the following formula for $q(t)$

$$q(t) = q_0(t) + \sum_{j=1}^{n} \xi_j(t) v^{(j)}(t), \tag{6}$$

which describes the time dependence of the state vector $q(t)$ through the evolution of the collective patterns (compare (2), (3)). Obviously, the reference state $q_0(t)$ can be called stable when all the possible excitations $\xi_j(t)$ decay during the course of time. When we now change the control parameters some (few) of the $\xi_j(t)$ can become unstable and start to grow in time. The border between decay and growth in parameter space is called a critical region.

As can be shown in detail (compare also Sect. 4) the few unstable amplitudes which we shall denote by u in the following change very slowly in the vicinity of a critical region whereas the damped modes s quickly decay to values which are completely prescribed by the unstable modes. It is this fact which is expressed by the slaving principle. The mathematical formulation reads [1, 2]

$$s = s(u)\,. \tag{7}$$

This relation allows us to eliminate the stable modes in (6) and leaves us with a low dimensional set of equations for the unstable modes which play the rôle of the order parameters. These equations then completely rule the behavior of the complex system on macroscopic scales near an instability.

The fundamental result of synergetics consists in the observation that on macroscopic scales new laws can be discovered which exist in their own right [1, 2]. These laws which are expressed by the order parameter equations turn out to be independent of the detailed nature of the subsystems and their interactions. As a consequence this allows us to introduce the concept of normal forms [7] as a method to discuss universal instabilities and qualitative dynamic behavior in the neighborhood of the critical regions.

3 The Method of Phenomenological Synergetics

As already noted, in discussing complex systems such as a society we run into the difficulty that we have incomplete knowledge about the subsystems and their interactions. Here the subsystems are individual humans, groups of humans, etc. This suggests two different methods to cope with such systems.

In the so–called *bottom up* approach the subsystems and their interactions are modelled by reasonable but quite simplified assumptions. The resulting equations are then treated along the lines which have been developed in mesoscopic synergetics [1, 2, 4–6]. The observation, however, that there are new universal laws on macroscopic scales suggests an additional method which may be called a *top down* approach [8]. Instead of constructing a simplified model on mesoscopic scales one starts directly from purely macroscopic considerations. The strategy then is as follows. One tries to identify macroscopic quantities from experience and classifies them according to time scale arguments. The slowest variables are usually identified with the control parameters which are assumed to be quasi static quantities. The slow macroscopic dynamics of the system has to be attributed to the order parameters. Very quickly relaxing variables have to be considered as enslaved modes (see Fig. 1).

It is important to note that there is no systematic strategy in social systems to identify these quantities properly. As we shall see, however, a careful use of experience may introduce new knowledge and new aspects to the understanding of nonlinear collective behavior in social systems. In the following section we work out suggestions for how it might be possible to apply synergetics by performing the identifications which were indicated above. This will be done by discussing some very simple examples.

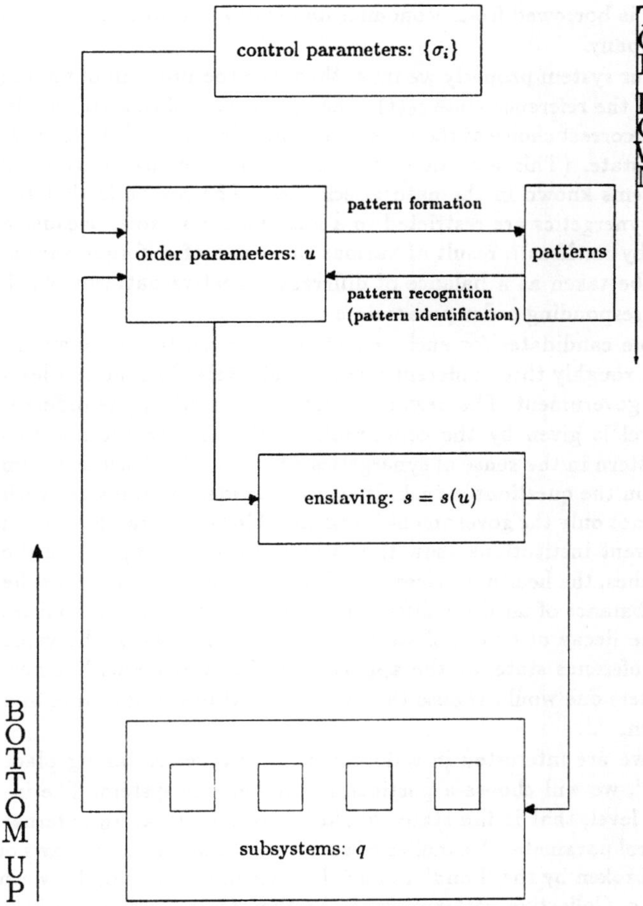

Fig. 1. Top down versus bottom up approach

4 Example of the Analysis of a Society

We shall introduce a method of constructing mathematical models in social systems on the basis of synergetics by using solely macroscopic observations. To be as concrete as possible we merely consider different structures which are met in the society of the Federal Republic of Germany. Our arguments can, however, be used in analyzing the relations between different states, economic problems, ecological systems, etc. After a general discussion of the problems which arise in identifying the reference state, the collective modes, the control and order parameters, we shall present two simple models. The first one is taken from the field of so-called psychosociology, i.e. the formation of public opinion. The second is borrowed from economics and is devoted to some aspects of the functioning of a company.

When we have chosen our system properly we must then face the problem of finding an exhaustive description of the reference state $q_0(t)$. The importance of this task lies in the fact that, for example, a correct choice of the collective modes (compare (6)) depends strongly on the reference state. (This assertion can already be seen by considering the simpler synergetic systems known in the natural sciences.) The reason is that the mathematical methods of synergetics are restricted to a local analysis. Now, because a society can be considered by itself as a result of various processes of selforganization, the reference state should be taken as a balance of different collective patterns which are associated with the corresponding order parameters.

We shall discuss possible candidates for such collective patterns. In Germany the government is organized on roughly three different hierarchical levels. The highest level is represented by the state government. The next lower level is formed by the different 'Länder'and the lowest level is given by the communities. We will characterize this structure as a collective pattern in the sense of synergetics which can be discussed from various aspects depending on the questions we ask. If we were to identify the state with our system we observe that not only the government is organized in this hierarchical form but nearly all socially relevant institutions show the same pattern. Examples are the trade unions, parties, churches, the health services, etc. The reference state can now be considered as a dynamical balance of all these different collective patterns. Instabilities are now connected with the decay of one of these patterns, the increase of the value of one pattern out of the reference state, or the appearance of new patterns like new parties. As control parameters one would choose the external relations of the state and its global economic situation.

If, on the other hand, we are interested in selforganization processes taking place on the level of the 'Länder', we will choose a particular 'Land'as our system. The influences of the next higher level, that is the state should be considered as an external constraint acting as a control parameter. Examples of such constraints would be laws of the state which cannot be broken by the 'Land', financial duties in the relation between the 'Land'and the state, etc. Collective patterns would be then the subpatterns of the collective modes which we identified on the level of the state.

We have mentioned these different situations in order to make the difficulties apparent which arise from the local nature of the theory. The careful definition of the "system"is necessary in order to proceed in its discussion from the synergetic point of view. Another difficulty which arises is how to weigh these different collective patterns by appropriately chosen variables. Furthermore, we note that in the present situation the tools which would enable us to give an exhaustive description of the reference state and the collective modes have not yet been developed in sociology [9]. A complete list of order parameters is therefore not available and only heuristic methods exist to determine some of them. As further important examples of order parameters we could mention the language, rituals, etc.

The result of the present discussion can therefore be described as follows. A given society can be considered as a macroscopic selforganized object. However, there are what might be called subprocesses of selforganization which can be discussed as phenomena existing in their own right. For these reasons it cannot be our concern to consider

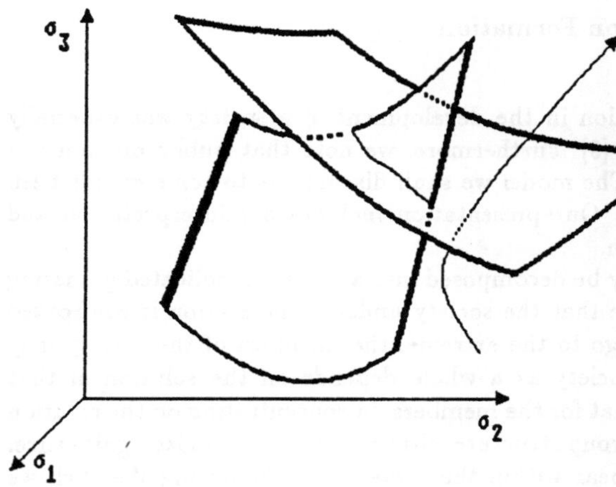

Fig. 2. Critical regions in parameter space

the evolution of a society by taking into account the interactions of all of these "order parameters". Instead of that ambition we are interested in the dynamics of such a society near well-defined simple instabilities. In these regions few new order parameters may emerge and rule the macroscopic dynamics in a comprehensible and simple fashion.

As already mentioned the way of reasoning outlined above can be applied to many other fields of scientific interest where collective behavior on the basis of selforganization is observed: In economics, ecology, population dynamics, etc. For this reason we discuss two different simple models from different social fields, the first in sociological and the second in economic terms. Both are universal in the sense mentioned above and have many further realizations in other fields. Using the simplest cases offers - at least in our opinion - the possibility of an interdisciplinary discussion of their value.

4.1 Preliminaries

Owing to the difficulties mentioned above we shall confine ourselves to two very simple examples of instabilities which can be observed in a society. The first simplification is the assumption that in both cases the control parameters of the open system are changed along a one dimensional manifold in parameter space (Fig. 2).

Instability regions in parameter space are assumed to have the properties of hypersurfaces (at least locally). When we now follow our one dimensional path through the parameter space we will typically meet only instability regions of codimension one. This is explained in Fig. 2 where we observe that the instability connected with an instability region of codimension one remain stable when we slightly disturb our path. In contrast, a critical region of a higher codimension will be removed by the slightest perturbation. This fact drastically reduces the types of possible instabilities.

4.2 Weidlich's Model of Opinion Formation

The important rôle of public opinion in the development of a society was especially emphasized by Lippmann in 1922 [9]. Furthermore, we note that public opinion can be measured by votes, polls, etc. The model we shall discuss has to some extent been proposed by Weidlich in 1971 [4]. Our presentation includes a reinterpretation and extension of its original formulation.

We consider a society which may be decomposed into a rather complicated gathering of groups, classes, etc. and imagine that the society under consideration is confronted with an important problem. If we go to the extreme, the members of the society may believe that the survival of the society as a whole depends on the solution of that problem. Experience now tells us that for the members, in concentrating on the solution of the given problem, the detailed group structure, citizenship, etc., loses its significance. Indeed, opposite opinions may appear within the same group. Following Weidlich we assume that only two solutions + and − are available and that there is no obvious preference for one of them. The decision of the society may then be measured by counting the numbers of votes which are given by n_+ and n_-, respectively. The problem may be further simplified by the assumption $n_+ + n_- = n$, where n stands for the total number of members and is assumed to be constant. Then just one macroscopic variable u can be identified which serves as an order parameter and which takes the form

$$u = \frac{1}{2}(n_+ - n_-) . \tag{8}$$

Taking n very large, we can consider u approximately as a continuous variable. We now have to seek the order parameter equation. Weidlich's hypothesis was to construct a probability density which is assumed to fulfill a master equation. The transition probabilities were taken to be isomorphic with the Ising model of a ferromagnet.

In contrast to that, we take a different point of view by stressing the rôle of the control parameter as a macroscopic quantity. Following Weidlich we choose his so-called adaption parameter σ. σ measures the influence of people on their neighbors. (We notice that it is by no means obvious how to attribute to σ in a given state of a society a precise value. We therefore can argue only qualitatively). It is well-known that there exists a whole class of equations which qualitatively yield the same situation. Mathematically this class can be precisely defined through the notion of topological equivalence. We are led to the conclusion that we can use the most simple representative of that class. This naturally yields the concept of normal forms. In the present case we obtain (neglecting the influences of the fluctuations)

$$\frac{d}{dt}u(t) = (\sigma - \sigma_c)u(t) - u(t)^3 . \tag{9}$$

Equation (9) has been discussed in detail in the article of Haken in this book. One observes a symmetry-breaking transition from a state of indifference to a state of polarization in a society. Connected with this transition are the well-known phenomena such as critical slowing down, critical fluctuations, etc. We want to emphasize that our way of reasoning is completely different from Weidlich's treatment. Indeed, he needed assumptions about the interactions of the members of the society to derive his moment equations which are of no relevance when one uses a macroscopic point of view.

We remark that when the macroscopic conditions for this model are realized, i.e. a change of one control parameter and symmetry between + and −, the interpretation of the mathematical form obtained appears realistic.

4.3 A Model from Economics

As a second example we consider a company and identify the control parameter as the amount of capital which constitutes its financial assets. For the company we identify as an important order parameter the profit obtained by producing goods or by providing a special service. We shall now study the nonlinear behavior of this order parameter when we change our control parameter.

Applying the method of normal forms [7] the only instability which can occur is the so-called saddle-node bifurcation. When we denote the order parameter by u its dynamics is described by the equation

$$\frac{d}{dt}u(t) = (\sigma - \sigma_c) - u(t)^2 \,. \tag{10}$$

Before we discuss the simple mathematical properties of this equation we want to emphasize what we have gained at this step. When we change one control parameter under the above assumptions the whole behavior of the complex system company is completely determined by such a simple equation of motion for the order parameter. It is important to note that (10) is a nonlinear equation which contains much more information than purely linear extrapolation could do. We substantiate this by first looking for the stationary solutions u_0 of this equation, which are characterized by a vanishing time derivative. They are given by the following formula

$$u_0^{\pm} = \pm\sqrt{\sigma - \sigma_c} \,. \tag{11}$$

We observe that there is no stationary solution if σ is smaller than σ_c. This result which rests on the nonlinearity of the equation can be interpreted in the following way. There is a minimal value of capital necessary so that the company can exist. Beyond σ_c two different solutions emerge where it turns out that u_0^+ is a stable solution (disturbances are damped) whereas u_0^- is an unstable solution (small deviations are enhanced). Again this has a simple interpretation. A company can only be stationary stable when it makes profits (positive value of the order parameter) and it will become unstable when it produces losses. The nonlinear effect here is concerned with the behavior in the neighborhood of σ_c. One observes that the stationary state crashes very fast with small changes in the control parameter (the slope of our curve goes to infinity).

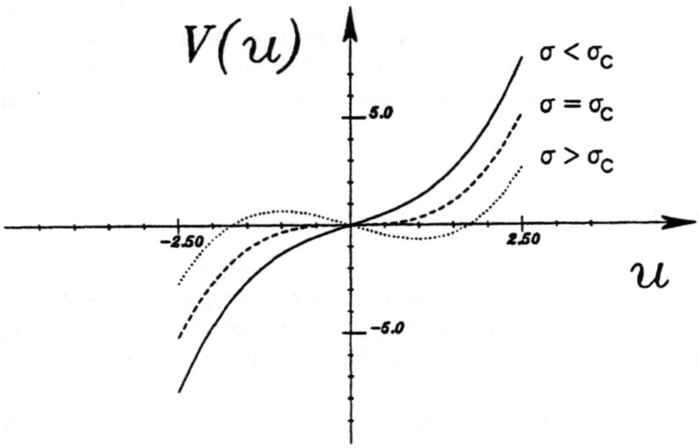

Fig. 3. Potential for the saddle–node bifurcation for different values of σ

To get some insight in the dynamical behavior we write (10) as

$$\frac{d}{dt}u(t) = -\frac{d}{du}V(u). \tag{12}$$

This allows us, following Haken, to interpret the dynamics of the order parameter in terms of the overdamped motion of a particle in the potential V which is given by

$$V(u) = -(\sigma - \sigma_c)u + \frac{1}{3}u^3, \tag{13}$$

where we put the arbitrary constant equal to zero. The potential V is drawn in Fig. 3 for three different values of σ.

In the case $\sigma < \sigma_c$ no minimum appears in the potential which means that the company will crash for any initial condition $u(0)$. In the case $\sigma = \sigma_c$ we have a turning point of the potential at the origin with slope zero. This indicates the appearance of a rest point at the origin. The smallest disturbance, however, still yields a crash of the company i.e. our particle escapes into $-\infty$. The situation changes dramatically when we consider the third case $\sigma > \sigma_c$. The stationary points of the bifurcation diagram are now represented by the minimum (u_0^+) and the maximum (u_0^-) of the potential curve. It remains interesting to note that not each initial condition of the order parameter leads to the stable minimum and a crash still remains possible. Again this has a simple and nice interpretation. Consider a situation where σ is smaller then σ_c and the particle is starting to escape to $-\infty$. If we were to avoid a crash we have to very quickly feed capital into the company in order to get a situation which is represented by our third case. The conclusion then is that it is still not always possible to avoid the crash. Again this is a property of the nonlinear behavior of the order parameter. We observe that our nonlinear form (10) does indeed reflect phenomena which also can be observed in reality.

5 Conclusion

By treating two simple examples we have proposed a method to understand the behavior of social systems on the basis of qualitative mathematical structures. The justification has been taken from the fundamental slaving principle which bridges the behavior of complex systems from the microscopic to the macroscopic level. The predictions we can give are only of qualitative value. Furthermore, they are restricted by the fact that they are only applicable locally in the state space Γ as well as in the parameter space $\{\sigma\}$. It is this fact that makes a precise reasoning on the basis of experience necessary in order to identify the control parameters, order parameters, etc. properly. However, we believe that a systematic study of social systems along the method outlined above will lead to new aspects in the understanding of social systems. Our simple examples were chosen to justify the method in the following way: We showed that a correct identification strategy led to mathematical structures which could be interpretated qualitatively and convincingly in the realm of the present results of social sciences. We take this as a justification in the sense that theory cannot be better than measurement.

Furthermore, our results shed new light on the problem of how to make predictions for the development of a society. Indeed, our method goes far beyond the results of linear extrapolation by explicitly taking into account the nonlinear behavior of the systems under consideration. Predictions are no longer unique, instead they are connected with well defined possibilities which are represented by universal mathematical structures and their unfoldings.

References

1. H. Haken: *Synergetics. An Introduction* (Springer, Berlin 1983)
2. H. Haken: *Advanced Synergetics* (Springer, Berlin 1983)
3. J. W. Forrester: *World Dynamics* (Wright–Allen, Cambridge 1971)
4. W. Weidlich: 'The Statistical Description of Polarization Phenomena in Society', Br. J. Math. Stat. Psychol. **24**, 51 (1971)
5. W. Weidlich and G. Haag: *Concepts and Models of a Quantitative Sociology* (Springer, Berlin 1983)
6. W. Weidlich: 'Physics and Social Science – The Approach of Synergetics', Phys. Rep. **204**, 1 (1991)

7. V. I. Arnold: *Geometrical Methods in the Theory of Ordinary Differential Equations* (Springer, New York 1983)
8. A. Wunderlin and H. Haken: 'Some Applications of Basic Ideas and Models of Synergetics to Sociology', in *Synergetics. From Microscopic to Macroscopic Order* (ed. E. Frehland), (Springer, Berlin 1984)
9. P. A. Sorokin: *Contemporary Sociological Theories* (Harper & Row, New York 1928)

Emergent Behavior in Insect Societies: Global Oscillations, Chaos and Computation

R. V. Solé[1], *O. Miramontes*[2], *and B. C. Goodwin*[2]

[1] Complex Systems Research Group, Dept. de Fisica i Enginyeria Nuclear, Universitat Politécnica de Catalunya, Pau Gargallo 5, 08028 Barcelona, Spain
[2] Dept. of Biology, The Open University, Walton Hall MK7 6AA, Milton Keynes, England

> *Achillles:* Familiar to me? what do you mean? I have never looked at an ant colony on anything but the ant level
>
> *Anteater:* Maybe not, but ant colonies are no different from brains in many respects ...
>
> Gödel, Escher, Bach; Douglas Hofstadter

Abstract: Insect societies are formed by a huge number of individuals in interaction. Ant behavior is simple and, apparently, predictable, but recent results suggest that low-dimensional chaotic dynamics would be implicated at the individual level dynamics. In this paper, we explore several recent experimental results concerning global properties of ant societies, with the individuals defined as chaotic automata

1 Introduction

Metamorphosis and flight were two revolutions in the evolutionary history of insects, together with the appearance of complex societies [1]. The social structure can be understood as a higher-level behavior of a set of simple and predictable organisms. Such units (say ants) are the basic structure of a "social mind" (the ant colony). The organization is maintained by interactions among many individuals, and two well known examples of such emergent structures are trail formation and nest building. Such behavior appears also in other systems in which some kind of computation is performed [2–4].

The analogies between ant colonies and brains are certainly not trivial. Both neurons and ants (Figs. 1a,b) have several common properties:

a. The behavior of single elements (ants or neurons) give us no information about how the colony or brain works: new phenomena arise when such elements are connected (in some way).
b. Global computational properties are observed as the outcome of the interaction of these structures and the environment.
c. When some number of elements are removed, the system is still able to show normal levels of performance [5–7].

The connection between the two structures has been pointed out several times in the past. In a 1969 book, Rémy Chauvin wrote: "We must also add that electronic engineers have now constructed circuits in which the different parts are joined to each other by

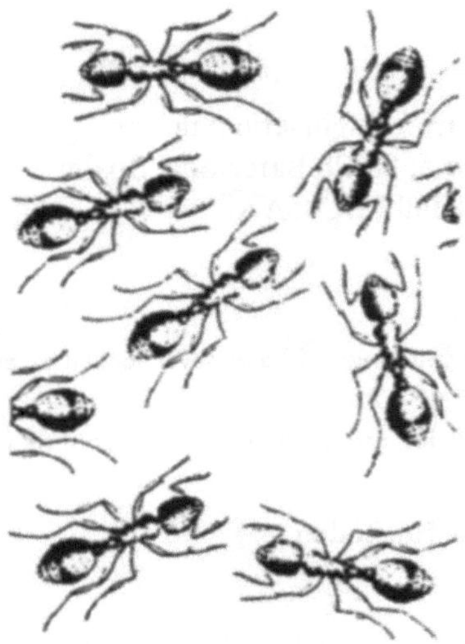

Fig. 1a. Neural organization of ant colonies. Interactions are not stable in time

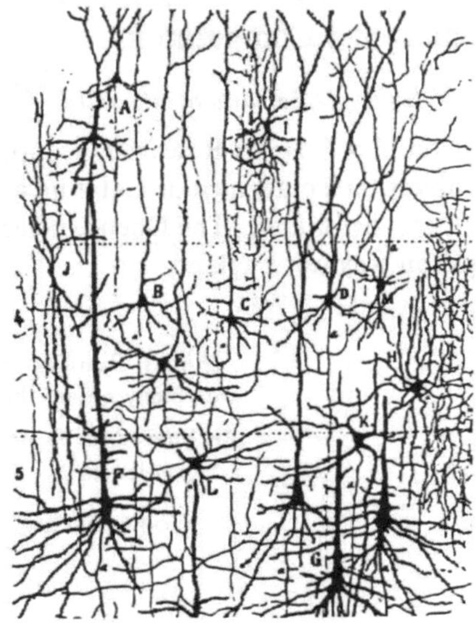

Fig. 1b. Neural network of human cortex (from Cajal, 1916)

as many connections as possible. If the connections are both numerous and random the whole network then has certain properties which remind one of the brain (...) I well know that an ant nest is not a brain, but this does not mean that the basic principles of their organization are not similar, or that the study of the ant colonies cannot teach us something about the brain, and vice versa. That would be a most unexpected result of the study of myrmecology..."

This suggestion is the core of our approach in this paper: several properties of neural systems can be applied (or translated) to ant colonies. Such properties can give us information about how both neural systems and insect societies behave.

These new properties are the result of the amplification of fluctuations in a synergetic system [2,8]. A key condition must be fulfilled: interactions must take place between nonlinear elements. As long as information transfer is linear, emergent behavior is absent.

Several experiments [6,9–11] clearly show that single individuals display more variable behavior. These differences between individuals are linked both to intrinsic physiological differences as well as new behavioral processes triggered by the individual's isolation. An example is search behavior (see below) in which individual ants behave apparently as random searchers. When they are put together, the emergent cooperative phenomena cannot be fully explained as the linear sum of individuals.

2 Neural Network Model

The study of real social insects [12] shows that the distribution of individuals among tasks is not a static phenomena. Ants interact at several levels (directly, through chemicals, etc) and this interaction can imply a change in the actual category (the internal state of a given ant) [6]. Some general points are well established:

a. Ant colonies are formed by a finite (usually large) set of individuals (i.e. simple units) which can be classified into some natural categories, by depending on the specific activity pattern. At a given moment (or over a time interval) each individual can be active or inactive [6,7].
b. Under normal (stationary) conditions, it can be shown that a given mean distribution of tasks remains more or less invariant. This special distribution is related to the fitness of the colony as a whole [13].
c. Perturbation experiments where the states of a given number of individuals (of some given classes) are externally changed shows that several categories responded non-linearly to these perturbations, in a way which suggests the existence of a complex process of parallel interaction [6].
c. The given task distribution of the ant colony is the result of a self-organization process emergent from local interactions.

The ant colony will be formally defined by a set of N individuals which are engaged into one of m given categories. The state of the i-th ant will be defined through the vector state $\mathbf{S}_i = \{S_i^1, ..., S_i^k\}$, where $S_i^j = \pm 1$ with $i, j = 1, ..., N$. With this state definition, there are $m = 2^k$ possible states available to each individual i.e. m possible "categories". A matrix structure is defined between these categories, i.e. Ω_{ij} with $i, j = 1, ..., m$. We consider a Boolean interaction rule: the p-th component of $\mathbf{S}_i$ will change under the effect of the p-th local field:

$$S_i^p(t+1) = \Phi\Big(\sum_{j=1}^{N} \Lambda_{ij}^p S_j^p(t)\Big) \quad ; \forall\ p = 1, ..., k \tag{1}$$

as in standard neural networks [5]. Here $\{\Lambda_{ij}^p\}$ represents the interaction strength associated with the p-th component. The couplings are defined by taking into account the actual categories: if $\mathbf{S}_i$ and $\mathbf{S}_j$ are individuals pertaining to the g and h categories respectively, then the interaction strength will be $\Lambda_{ij}^p = \Omega_{hg}$.

The couplings are then determined in a simple way from the current states. As a given individual shows a category change, the corresponding couplings also change.

Several choices of Φ are possible; we take in our model $\Phi(z) = sign(z)$. An energy function can also be defined in the usual way:

$$H(\{\Lambda_{ij}^p\}, \{S_j^p\}) = -\frac{1}{2}\sum_{p=1}^{k}\sum_{i,j=1}^{N} \Lambda_{ij}^p S_i^p S_j^p. \tag{2}$$

and the corresponding stability condition is described by the inequality

$$\{S_i^p h_i^p > 0\} \quad \forall i = 1, ..., N \ ; \forall p = 1, ..., k$$

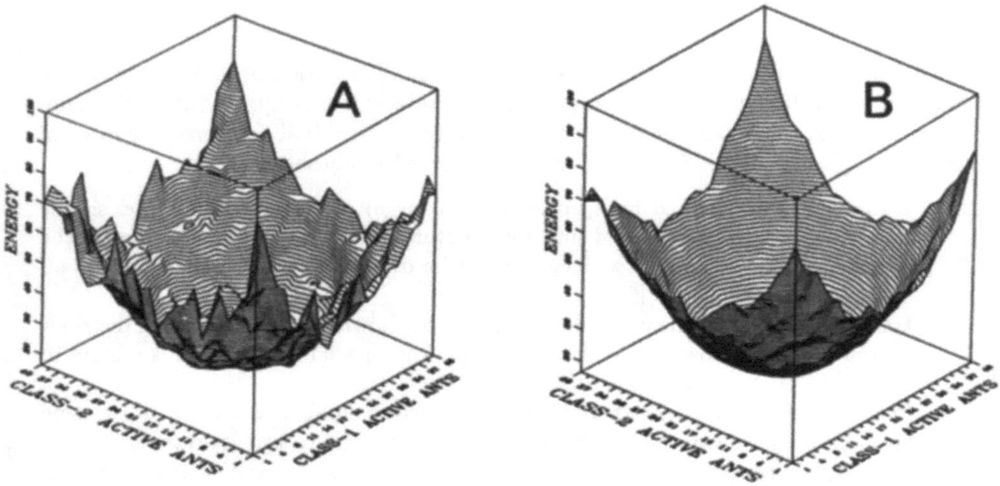

Fig. 2. Energy surfaces for a n=80 colony with 8 categories. (a) from one single sample and (b) averaging over 10 samples (see text)

h_i^p being the local field i.e.

$$h_i^p(t) = \sum_{j=1}^{N} \Lambda_{ij}^p S_j^p(t). \tag{3}$$

Here two restrictions on the matrix coefficients are imposed: the elements are symmetric i.e. $\Lambda_{ij}^p = \Lambda_{ji}^p$ and there is no self-interaction i.e. $\Lambda_{ii}^p = 0$ [5].

In field studies [6] it is possible to determine the actual category of a given ant, and we can eventually define the current state of the ant colony by: $\mathbf{S} = (\mathbf{S}_1, \mathbf{S}_2, ..., \mathbf{S}_N)$ but a more useful quantity for our purposes is $\mathbf{n} = (n_1, n_2, ..., n_m)$ which gives us the number of individuals in each category at a given time step (numbers of foragers, patrollers, etc). Under unperturbed conditions, we can consider that the colony behavior is defined by a given reference state, say $\mathbf{n} = \mathbf{n}_0$. With this description, the energy function will be (for $n_k \gg 1$; $\forall k = 1, ...m$):

$$H(\{\Lambda_{ij}^p\}, \mathbf{n}) = -\frac{1}{2} \sum_{l=1}^{m} \sum_{r=1}^{m} C_{lr} n_l n_r. \tag{4}$$

In order to illustrate this result, let us consider a trivial situation defined by a system of N spins with two defined categories: up (+1) and down (−1). If N_1 and N_2 are the corresponding numbers for each category we have $N = N_1 + N_2$, and we will consider that the interaction between the two types are not equivalent. In this sense (and this is a key difference) the effective interactions change with each spin flip. The field associated with a given i-th up spin is: $h_i = \sum_j^N = (N_1 - 1)J_{11} - N_2 J_{12}$ being J_{pq} the interaction strength between a (p, q)-pair. For a down spin the local field will be $h_j = N_1 J_{21} - (N_2 - 1)J_{22}$. For $N_1, N_2 \gg 1$ and using $J_{11} = J_{22} = \beta$ and $J_{12} = J_{12} = \alpha$, we have $h_i(+1) = \alpha N_1 - \beta N_2$, $h_j(-1) = \beta N_1 - \alpha N_2$. Finally, the corresponding energy function will be:

$$H = -\frac{1}{2}\sum_{i,j} J_{ij}S_iS_j = -\frac{1}{2}\sum_i S_i\Big(\sum_j J_{ij}S_j\Big) = -\frac{1}{2}\sum_i S_ih_i$$

$$= -\frac{1}{2}\left[\sum_{S_i=+1} S_ih_i + \sum_{S_j=-1} S_jh_j\right] = -\frac{1}{2}(\alpha N_1^2 + \alpha N_2^2 - 2\beta N_1N_2) = -\frac{1}{2}\sum_{k=1}^{2}\sum_{l=1}^{2} C_{kl}N_kN_l$$

with $C_{11} = C_{22} = \alpha$ and $C_{12} = C_{21} = -\beta$.

Using this new definition we can describe the energy function in the m-dimensional space defined by $(n_1, ..., n_m)$. An image of the energy surface (see Fig.2) can be obtained by representing a two-dimensional section given by the surface $H(n_1, n_2; R\{n_{k>2}\})$ in which H is calculated using the pairs of points (n_1, n_2) linked with the two first categories (here $1 \leq n_1, n_2 \leq N/2$) and where the other $N' = N - n_1 - n_2$ individuals are randomly selected from the other categories.

If we only use a sample of such combinations, a rugged surface is obtained; using ten samples and averaging, we obtain a smooth surface. In our example (see ref [7]) we have an attractor at $\mathbf{n}_0 = (20, 20, ..., 20)$. The bottom of the energy landscape at our section surface is observed at $(20, 20)$. It can be shown that the colony movement towards $\mathbf{n}_0$ is robust with respect of local interactions. As a consequence of our previous results, we can understand how the task distribution can move in phase space after perturbation experiments. The nonlinear character of interactions causes changes in several tasks because of the existence of cooperative phenomena. These changes cannot be explained as a linear sum of isolated events.

The previous problem deals with the existence of a stable point attractor. Such an attractor can be interpreted, in adaptative terms, as the best distribution of workers and steady state seems to be a typical situation. However, oscillations and chaos in the numbers of active workers have been recently described [10,11,14]. In the next section, we will analyse such situation by using a cellular automaton (CA) model.

3 Periodic Oscillations with Chaotic Elements

Ant colonies are able to show a very unexpected kind of dynamical behavior: global oscillations of activity levels [10]. This means that, under some conditions, some ant colonies show short-time periodic changes in the numbers of active workers. Recent studies [6,10,11] have revealed the existence of nonlinear dynamics through local interactions.

One of the results of Cole's experiments was that single ants behave as chaotic elements. He measured the fractal dimension D_f of single ant dynamics and an estimation over five samples gave $< D_f > \approx 2.43$, corresponding to that of a strange attractor. When global dynamics was analysed, the value over the same number of samples was $< D_f > \approx 3.09$, linked with periodic motion.

This result, obtained from studies in laboratory colonies of ***Leptothorax allardycei*** is in fact the first experimental evidence of chaotic dynamics in animal behavioral processes. Chaos would serve as a source of unpredictability and flexibility in search and

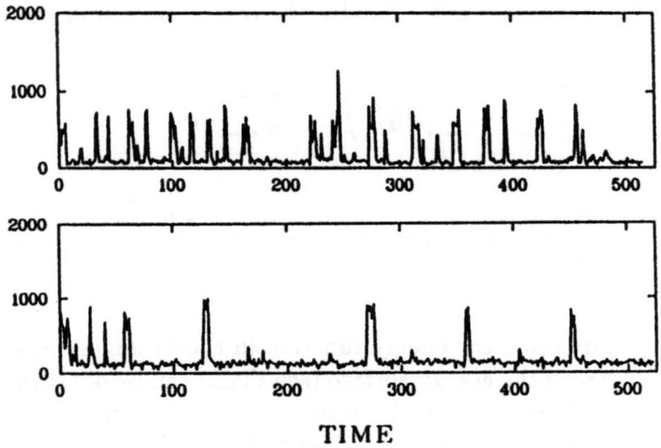

Fig. 3. Individual oscillations (From Cole [10])

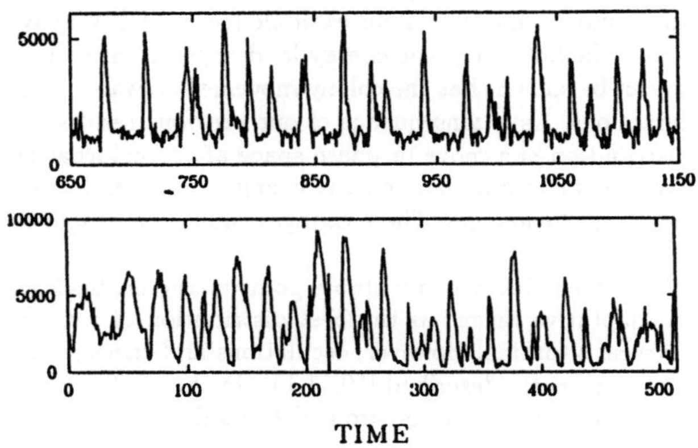

Fig. 4. Global coherent oscillations (From Cole [10])

predator avoidance. For a neural system with simple organization, low dimensional deterministic chaos represents the simplest way in which a nonlinear selforganized system can show randomness. In fact, the same mechanisms and structures implicated in other phenomena in which steady states are required, can generate chaos when some bifurcation parameter is changed above a given threshold. External inputs (or their absence) can be the source of such bifurcations.

Let $\Xi(n,t) = \{S_1(t), ..., S_n(t)\}$ be the global state of a colony of n individuals (all microscopic states). Here $S_i(t) \in \mathbf{R}$ is the activity state of the i-th individual. A given ant is called active if $S_i(t) > \theta$ and inactive otherwise. Here θ is a threshold below which the ant becomes unactive (here we take $\theta = 0$. In order to compare our results with experimental data, we study the number of active elements, $0 \leq n_a(t) \leq n$.

Movement rules are simply stated. If a given ant is active, each time step a deplacement towards an empty neighboring point will occur. The movement rule is then defined as $S_k(\mathbf{r};t) \to S_k(\mathbf{r}';t+1)$. If the ant is inactive or all nearest points are occupied, no movement takes place.

Let $C(\mathbf{r}) \subset \Lambda(L)$ be a neighborhood of a given ant $S_k(\mathbf{r};t)$. In our study, only the eight nearest lattice points are considered. Then the dynamics of ant states will be described by:

$$S_k(\mathbf{r};t+1) = tanh\Big[g \sum_{\mathbf{q}\in C(\mathbf{r})} J_{km} S_m(\mathbf{q};t)\Big] \qquad (5).$$

The set of matrix elements $\{J_{km}\}$ defines the specific kind of interaction and dynamics. For simplicity, we take in our study $J_{km} \in \{-1,0,+1\}$ and several matrices are used. A further and nontrivial assumption is also introduced: self-interaction of individuals is present. In Cole's experiments, isolated ants were studied on a two-dimensional domain. In macroscopic terms, the previous observed behavioural patterns are in fact a kind of excitable medium [15,16]. Each activation event gives birth to a wave of excitation throughout the colony. In mathematical terms, this situation resembles of that an excitatory neural tissue [16]. Let $a(x,t)$ the activity level at a given spatial position x. The evolution of the activity can in fact be approximated by:

$$\frac{\partial a}{\partial t} = -\gamma a + \epsilon \int\int_{\Gamma} e^{(-\delta|x-x'|)}\, tanh(ga)\, dx' + R(a-\theta) \qquad (6)$$

with $\delta, \epsilon, \gamma > 0$ being some given constants and R a threshold-dependent function including random pulses (activations). Under some conditions, this model will be able to develop oscillations of activity, but using this approach makes difficult a careful comparison with the previous experimental results, in particular those results dealing with the number of individuals in interaction.

As a microscopic model, we consider here the dynamics of individual ants as defined by a so called mobile cellular automaton (MCA) [15,17], being each ant described as a nonlinear dynamical system.

4 Strange Attractors, Oscillations and Spatial Structure

The chaotic dynamics of activation in the observed ant colonies strongly suggest the existence of a low-dimensional chaotic mechanism in the nervous activity of individual ants. Such a mechanism will act at the level of activation phenomena, and will be modelled here by using a continuous chaotic system. Here we will use the following neural model (Ermentrout, 1984):

$$\frac{dx}{dt} = y - \mu x \;;\; \frac{dy}{dt} = z - \nu x \;;\; \frac{dz}{dt} = qx^2 - \gamma x. \qquad (7)$$

The stability analysis of the rest equilibrium point $\mathbf{x}_0 = (0,0,0)$ shows that this point is stable iff $\mu, \nu, \gamma > 0$ and $\mu\nu - \gamma > 0$ (Routh criterion). If $\gamma = \mu\nu$, the eigenvalues are $\lambda_1 = -\mu$ and $\lambda_{2,3} = \pm i\sqrt{\nu}$ and for $|\gamma - \mu\nu|$ small, a limit cycle can emerge through Hopf bifurcation. Using $\mu = 1, q = 2$ and $\nu = 2$, the first bifurcation occurs for $\gamma_c = 2$ and this limit cycle becomes unstable for $\gamma \approx 3.09$, the first period-doubling bifurcation. For $\gamma > 3.45$, chaotic attractors are present, as shown in Fig. 5.

Here we take $z(t)$ as the "control variable" which determines the activation sequence of individuals: an inactive individual will become active if $z(t) - \theta > 0$, where θ is a

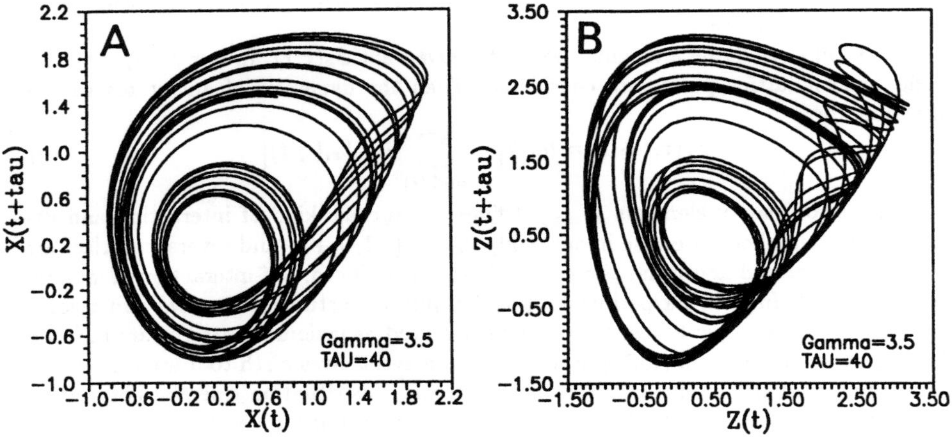

Fig. 5. Strange attractors obtained from eq. (7) for $\gamma = 3.50$. Two reconstructions have been used: (a) $x(t) - x(t + \tau)$ with $\tau = 40$ and (b) same as before with z(t)

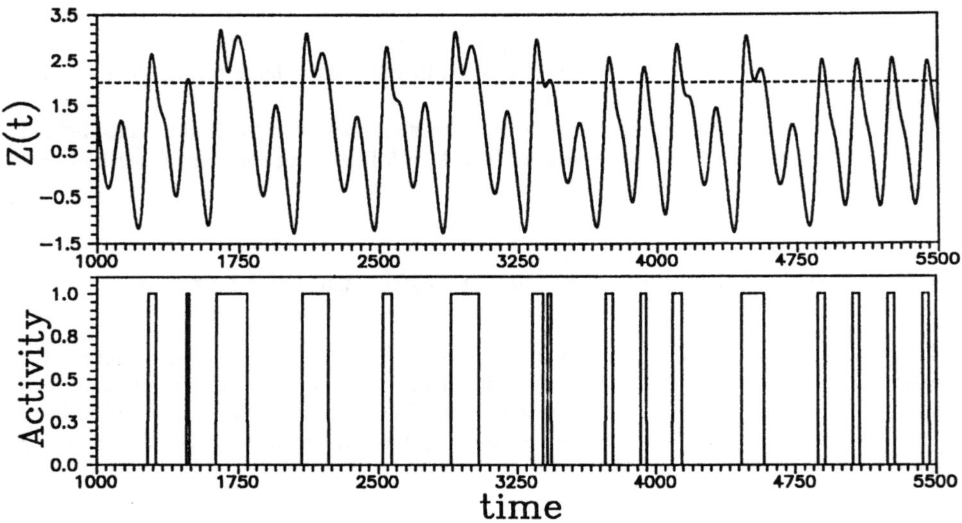

Fig. 6. Nonlinear dynamics of a single individual. We show: (for $\gamma = 3.50$ and $\theta = 2.0$) (a) $z(t)$ (from eq.7) and (b) activation state ; here $s_i = 1$ if activated and $S_i = 0$ otherwise.

given threshold. In Fig. 6, an example of the $z(t)$-dynamics is shown together with the activation state (1=active, 0=inactive) for a single ant.

Now we consider the behavior of a set of n interacting individuals. The interaction matrix (following the previous approach) is defined for active-inactive ants:

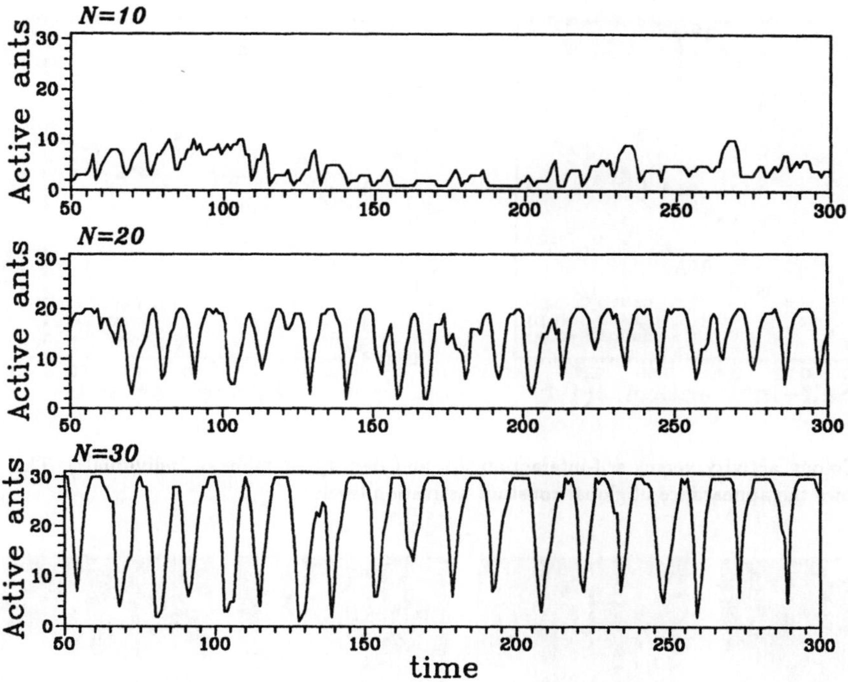

Fig. 7. Global oscillations from ant colony dynamics (model)

$$J = \begin{pmatrix} J_{11} & J_{12} \\ J_{21} & J_{22} \end{pmatrix}$$

in which: J_{11}: active-active coupling (now self-interaction is considered), J_{22}: inactive-inactive coupling and J_{12}, J_{21}: active-inactive coupling. Equivalently, we can write J as $J = \{J_{11}, J_{12}, J_{21}, J_{22}\}$. Several matrices have been used in our study. In the following we take one of the simplest ones i.e. $J = \{1, 1, 1, 0\}$ as a model of interaction.

Those matrices with $J_{11} = 0$ seem to be unlikely to show coherent oscillations. We also take $J_{22} = 0$ i.e. no-interaction between inactive individuals (equivalent results are obtained for non-zero value). The cross-terms i.e. $J_{ij}, i \neq j$ can have different values, and have been found to be less important in generating collective oscillations. The frequency and amplitude of such oscillations can, however, be changed by depending on these cross-interactions.

Now, using our set of automata under the previously described dynamics (eq.5), we find that, by increasing the number of individuals, periodic oscillations are present as shown in Fig.7. Increasing the colony density (i.e. $\rho = N/L^2$) with L=7, we can see that the global pattern of activation is periodic for enough high densities, as observed in experimental colonies [10,14]. The density is clearly related to the existence of percolation of activity. The computation of the Fourier spectrum shows the emergence of a well defined dominant peak as ρ is increased. In particular, we can analyse the effect

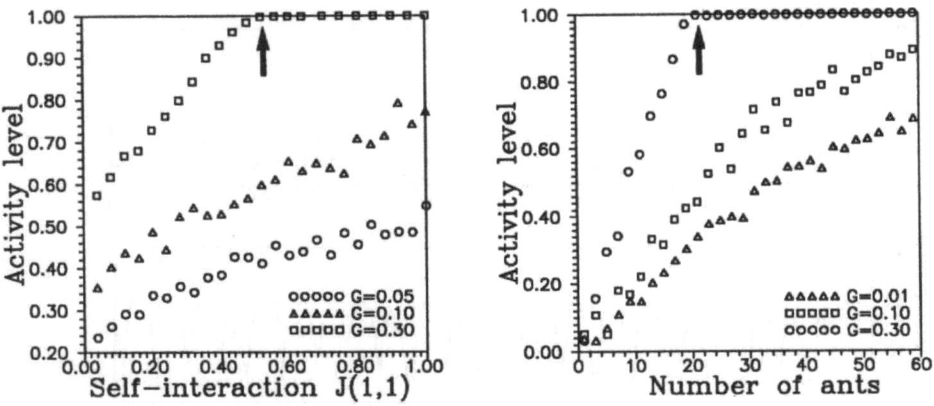

Fig. 8. Colony activity versus self-interaction J_{11} and density (number of individuals). The arrow shows the appearance of global constant activation levels

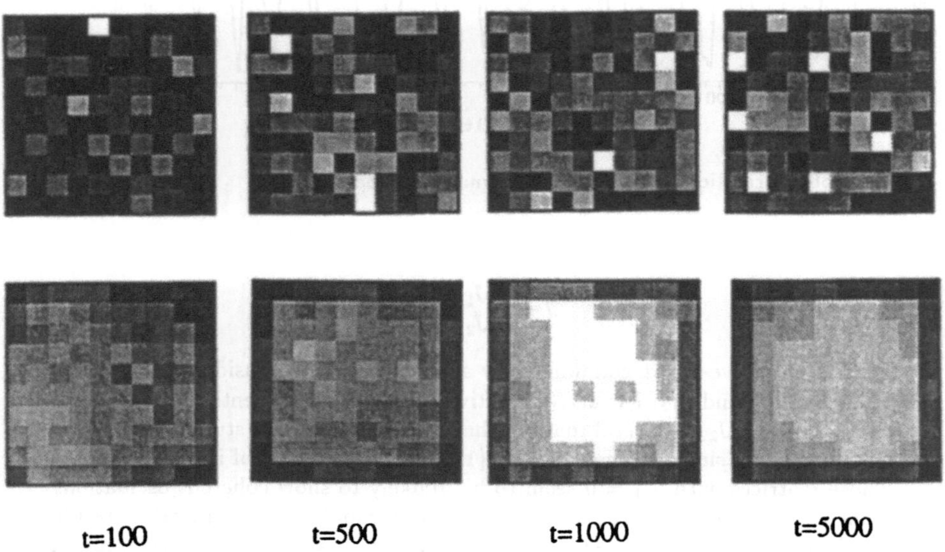

Fig. 9. Spatiotemporal distribution of activity. Top:no local interaction between individuals. Bottom: local interaction with nearest lattice points

of several parameters on the periodicity and numbers of active elements. In Fig.8 the activity level (the number of active ants at each point averaged over 500 time steps) is shown as a function of the self-interaction and the number of individuals. As we can see, higher J_{11}-values give higher activity levels with saturation beyond some threshold (see arrows). Still more interesting, the effect of density is also clear from Fig. 8b. In both cases, we can see that the activity and as a consequence the use of energy resources is strongly dependent on both parameters.

Here we can speculate about the possible adaptative meaning of these oscillations: the colony can operate at a nonstationary state (as in some physiological systems) being able to minimize the use of energy (enough to guarantee brood care and colony survival). As the number of active units increases, the energy is increasingly used, and such situation can move to a full-time activated colony. The increasing levels of activity could be able to trigger colony fision, as observed in experimental colonies. Finally, this activity pattern is reflected in the spatial distribution of activity. A recent study using this model with random activation [15,18] has shown that activity is distributed in concentric patterns (Fig. 9) if self-interaction is present. This result also explains the selforganizing pattern of the brood observed inside the ant nests, also distributed as concentric rings [19]. Once more, the introduction of local interactions in a spatially distributed system results in new emergent phenomena [20,21].

5 Discussion

Emergent behavior is shown to be present in insect societies. These emergent properties are the outcome of synergetic processes obtained from the nonlinear interactions. In our study, global oscillations are present as the macroscopic pattern of spontaneous activation. The individual behavior was described through a chaotic attractor; using a neural-like interaction, the global pattern of activity acts as an order parameter which "enslaves" the individual behavior of ants. When isolated, these individuals are again random-like units. Such individual states can be used as an adaptation as shown by Deneubourg [22-24]. Low dimensional chaos can provide the (deterministic) source for such probabilistic behavior [22]. Oscillations are able to synchronize the ant colony as a single macroscopic unit. The spatial patterns of activity can be the source of order, and again the observed brood distribution inside the ant nests is here obtained from the emergent behavior.

It is well known that several properties of insect societies can be explained on a genetic basis [1,13]. Are then nonlinear phenomena and synergetics necessary in our understanding of evolution or social behavior?. Genetic constraints are, we believe, only part of the story. In our examples, the local structure of interactions (which would be encoded at the genetic level) gives birth to higher-order phenomena (which are not). In this context, the emergence of a social structure as the ant colony can be seen (to some extent) as the result of a search in the "space of emergent properties". A full understanding of these systems needs a complementary approach from both levels of observation.

Ackowledgments

The authors would like to thank Profs. H.Haken and A.Mikhailov for several useful comments as well as Jordi Bascompte for discussions on evolution and social behavior. This work has been supported by grants of Universitat Politécnica de Catalunya, UPC PR9119,and CIRIT EE91/1 (Ricard V. Solé) and of Universidad Nacional Autónoma de México and British Council (Octavio Miramontes).

References

[1] Wilson, E.O., (1972) The Insect Societies, Harvard U. Press.

[2] Haken, H. and Stadler, M., eds. (1990) Synergetics of Cognition. Springer-Verlag (Berlin)

[3] Kephart, J.O., Hogg, T. and Hubermann, B.A. (1990) Collective behavior of predictive agents. Physica D42, 48-65.

[4] Kiss,G. (1991) Autonomous agents, AI and Chaos theory. in:*From animals to animats.* The MIT Press (CAmbridge, USA).

[5] Amit, D.J. (1989):" Modelling Brain Function". Cambridge University Press.

[6] Gordon, D.M. (1989) Dynamics of task switching in harvester ants. Anim. Behav. 38,194-204.

[7] Gordon, D.M., Goodwin, B.C. and Trainor, L.E.H. (1992) A Parallel Distributed Model of the Behavior of Ant Societies. J.Theor.Biol. 156, 293-307.

[8] Nicolis, G. and Prigogine, I. (1977) Selforganization in Nonequilibrium Systems. Wiley and Sons, New York

[9] Chen,S.C. (1937): Social modification of the activity of ants in nest-building. Physiol.Zool. 10,420.

[10] Cole, B.J. (1991a) Short-term activity cycles in ants: generation of periodicity by worker interaction. Am.Nat. 137,244-259

[11] Cole, B.J. (1991b) Is animal behavior chaotic? Evidence from the activity of ants. Proc.Royal Soc.Lond. B244, 253-259.

[12] Gordon, D.M. (1989b) Caste and change in social insects. In *Oxford Surveys in Evolutionary Biology* 6. (Harvey, P. and Partridge, L., eds.). Oxford U. Press.

[13] Oster,G. and Wilson, E.O. (1978) Caste and Ecology in the Social Insects. Princeton U. Press.

[14] Franks, N.R., Bryant, S., Griffiths, R. and Hemerik, L. (1990). Synchronization of the behavior within nests of the ant *Leptothorax acervorum* I. Bull.Math.Biol. 52,597-612.

[15] Miramontes, O., Solé, R.V. and Goodwin, B.C. (1992) Collective Behavior in random-activated Cellular Automata, to appear in Physica D.

[16] Mikhailov, A.S. (1990) Foundations of Synergetics I, Springer Series in Synergetics, Springer (Berlin).

[17] Solé, R.V. , Miramontes,O. and Goodwin,B.C. (1992) Oscillations and Chaos in Ant Societies. To appear in J.Theor.Biol.

[18] Miramontes, O. (1992) Complexity and behaviour in Leptothorax ants. Mh D Dissertation.

[19] Franks, N.R., Sendovafranks, A.B. (1990). Brood Sorting by ants: distributing the work load over the work surface. Behav. Ecol. Sociobiol. 30, 109-119.

[20] Solé, R.V. and Valls, J. (1992) On structural stability and Chaos in Biological Systems. J Theor.Biol. 155,87-102.

[21] Solé, R.V. , Bascompte, J. and Valls, J. (1992) Nonequilibrium Dynamics in Lattice Ecosystems: Chaotic Stability and Dissipative Structures. Chaos 3,387-397.

[22] Conrad, M. (1986)in: Chaos (A.V.Holden, ed.); Manchester Univ.Press.

[23] Deneubourg, J.L., Pasteels, J.M. and Verhaeghe, J.C. (1983) Probabilistic behavior in ants: a strategy of errors?. J.Theor.Biol. 105,259-271.

[24] Deneubourg, J.L., Goss, S., Pasteels, J.M. and Duerinck, G. (1986) Random behavior, amplification processes and number of participants: how they contribute to the foraging properties of ants. Physica D22, 176-186.

Collective Dynamics in Models of Communicating Populations

A. S. Mikhailov

Abteilung Physikalische Chemie,
Fritz-Haber-Institut der Max-Planck-Gesellschaft,
Faradayweg 4–6, W-1000 Berlin 33, Fed. Rep. of Germany

Abstract. Three characteristic examples of the populations employing different modes of communication, such as exchange of addressed messages, mass communication and remote sensing, are investigated. We find that the populations are able to collectively perform some functions of the information processing that are typical for neural networks.

1. Introduction

The living beings do not passively obey the physical forces coming from the environment. They are able to maintain autonomous activity, to change their internal states and to steer their own motions. Therefore, the role of the environment consists rather in providing the *signals* which are further processed by the biological organisms and trigger their specific responses. The interactions between these organisms are based on generation, reception and processing of signals, i.e. on various forms of *communication* between them.

The collective dynamics of a communicating biological population is determined by the rules of communication between the individual elements and the repertory of their functional responses to the received signals. Given the great variety of biological species, it is hardly possible to develop any universal theory of communicating populations. A more fruitful approach might consist in construction and investigation of simple abstract models that do not describe a particular biological system but can shed light on the relationship between the possible collective dynamics of a population and the set of the basic communication rules which it employes.

Obviously, the theoretical study of communicating populations should start with the simplest cases, where both the rules of communication and the functional responses are well defined. A good opportunity for this is provided by the biological systems which use the chemical form of communication.

Chemical signals play an important role in organization of behaviour in cell populations, as demonstrated by the studies of the slime mould

Springer Series in Synergetics, Vol. 62 **Interdisciplinary Approaches to Nonlinear Complex Systems** – Eds.: H. Haken and A. Mikhailov

Dictyostelium discoideum [1,2] and other microbial systems [3,4]. The neurons of the brain are known to release a large number of different neuromediators which influence the behaviour of other neural cells [5] and should contribute into the emergent properties of the information processing. Even the biological cells which form single living plants can chemically communicate because, as it was recently found [6], they are connected by a network of transmembrane channels that allow the transport of macromolecules from one cell to another.

Communication between the biological macroorganisms is also often chemical. This is especially true for the insects which leave in the medium the chemical traces that are sensed by other members of the population and determine their functional responses. In the insect societies, formed by ants, bees etc., such exchange of chemical signals results in very complex patterns of the collective behaviour [7].

Note that, although the higher animals and humans employ much more sophisticated communication language, the primitive aspects of their communication-induced collective behaviour can be sometimes described by the mathematical models which are not much different from those dealing with the chemical communication.

Recently, much attention has been paid to the theoretical studies of cell-to-cell signalling in populations of *Dictyostelium discoideum* (see, e.g., [8,9]). These collective amoebas can sense presence of the concentration gradients of a certain chemical substance and can respond by generation of the same substance and its release into the medium and by active motion along the direction of the gradient. This results in emergence of self-sustained wave-like motions and formation of complex spatio-temporal patterns. The mathematical models of these communication-induced processes are similar to the models of chemical excitable media, such as the famous Belousov-Zhabotinskii reaction [10].

In the present paper we examine a different class of models. In contrast to the studies of pattern formation in *Dictyostelium discoideium* and related systems, where all individual elements are assumed to be functionally identical, we want to investigate the properties of collective dynamics in the populations of the *differentiated* individual active elements. In our models the elements keep their individual rules of response and generation of the chemical signals. We find that in some aspects the collective dynamics of such populations can approach the kinds of collective behaviour which are involved in the currently discussed models [11,12] of distributed information processing. It gives evidence that communication between the individual members of a population can give rise to emergence of a form of a collective intelligence.

2. Addressed Messages

About a decade ago Shepherd [5] suggested that exchange by chemical signals can effectively lead to the same sort of informatic interactions that are realized by direct synaptic connections between the neurons of the brain. As is shown below (following [13,14]), it is indeed possible to formulate the laws of communication in such a way that *any* pattern of activity of a neural network is effectively reproduced by a communicating population.

Suppose that the members of a population (i.e. individual cells or macroorganisms) can communicate by sending and receiving *addressed messages* (which may be materially carried by appropriate chemical molecules). To simplify the model, we assume that any individual member j of the population has only distinct functional states ("up" and "down") which are specified by a binary variable S_j that takes values $S_j = +1$ ("up") and $S_j = -1$ ("down"). Any given message consists of two parts: it bears an address tag, that indicates to which population member j it is sent, and the proper message which is encoded in the binary variable σ that takes values $\sigma = \pm 1$. A message (σ, j) is an *order* addressed to the member j; if $\sigma = +1$ it dictates that this element should go into the "up" state ($S_j = +1$), if $\sigma = -1$ the element is required to go to the "down" state.

Within a small time interval, any member of the population can receive many different messages with contradicting orders. Its actual response is determined then by the dominant order, which is contained in the greater number of the messages. It means that the new state S'_j of the element j is

$$S'_j = \mathrm{sign}(m_{+1,j} - m_{-1,j}) \quad , \tag{1}$$

where $m_{\sigma,j}$ is the number of the received messages (σ, j).

Next, the rules of generation of messages should be specified. We assume that any member i of the population is permanently sending messages to *all* other members j. The *frequency* I_{ij} of generation of these messages depends on addresses i and j of the sender and the receiver but it is independent of the current states of these elements. However, the *order* σ transmitted by a message depends essentially on the current state of the sending population member: if the state of the sender is reversed, the same happens with the order which is being sent.

The orders sent by an element do not generally coincide with its momentary state. Instead, each element i *knows* which order $\sigma = w_{ij}$ it should send to any other element j when the first element is in the "up" state (i.e. if $S_i = +1$). If the element i is currently in the opposite state $S_i = -1$, it sends the order $\sigma = -w_{ij}$. Note that thus we consider a population of the differentiated elements whose responses are individual, i.e. determined by the identification address of an element.

We assume that the messages are released into the medium where they can reach a receiver without any significant delay. Once released into the medium, the messages have a finite life-time τ (this prevents their accumulation in the medium).

The time evolution of the mean number $m_{\sigma,j}$ of messages (σ,j) obeys a kinetic equation

$$\dot{m}_{\sigma,j} = - m_{\sigma,j}/\tau + \sum_i I_{ij} (1 + \sigma w_{ij} S_i)/2 \quad . \tag{2}$$

It can be easily seen that an element i gives a nonvanishing contribution into the generation rate of the messages (σ,j) only if $\sigma = w_{ij}$ for $S_i = +1$ and if $\sigma = -w_{ij}$ for $S_i = -1$.

If an average time interval between the changes in the states of the elements is large as compared with the characteristic life-time τ of the messages, the numbers of the messages adjust to the momentary states of the active elements and we have

$$m_{\sigma,j} = (\tau/2) \sum_i I_{ij}(1+ \sigma w_{ij} S_j) \quad . \tag{3}$$

Hence, the difference $h_j = m_{+1,j} - m_{-1,j}$ in the numbers of the opposite orders received by an element j is

$$h_j = \tau \sum_i I_{ij} w_{ij} S_i \quad . \tag{4}$$

According to (1), the new state S'_j of an element j is determined then by the sign of h_j, i.e.

$$S'_j = \text{sign } h_j \; . \tag{5}$$

It is interesting now to compare the resulting collective dynamics of such population with the functional behaviour of neural networks.

In the simplest mathematical model of a neural network, which was introduced by McCullloch and Pitts [15] in 1943, each neuron j is considered as an element with only two possible state $S_j = \pm 1$. When $S_j = +1$, the neuron is in the "active" state, while $S_j = -1$ corresponds to the "passive" state of a neuron. Each neuron is assumed to be physically connected to all other neurons. The connections are characterized by a "synaptic weights" J_{ij} that can take both positive and negative values (in the latter case the connection is called inhibitory). At randomly chosen time moments the neuron update their current states. The new state S'_j

is determined by the rule S_j = sign h_j where the acting field h_j consists of the sum of the contributions from all other neurons,

$$h_j = \sum_{i \neq j} J_{ij} S_i \quad . \tag{6}$$

Despite the apparent simplicity of this mathematical model, it can support very complicated forms of dynamics. With an appropriate choice of the synaptic matrix J_{ij}, it can demonstrate the properties of associative memory or store and retrieve the temporal sequences of the activity patterns. There are also many effective learning algorithms for this model, based on the gradual adjustment of the synaptic weights.

Comparing (4) and (6), we see that our model of a communicating population is mathematically equivalent to this formal model of a neural network. We see that a neural network with an *arbitrary* synaptic matrix J_{ij} can be implemented by the communicating population if we choose

$$I_{ij} = (1/\tau)|J_{ij}| \quad , w_{ij} = \text{sign } J_{ij} \quad . \tag{7}$$

Thus, our analysis confirms the suggestion by Shepherd [5] that communication can provide effective "distant synapses". The emergent properties of the population are identical to those of a neural network. The communicating populations can thus perform the same tasks of information processing and learn from experience in the same way as neural networks.

The required amount of communication in the population is, however, very large. To imitate the synaptic connections, its members must permamently generate messages which are addressed individually to all other members of the population. If the communication is based on release of different chemical molecules, it means that the number of employed molecular molecules should be twice larger than the number of the elements in the population (because, in addition to an address, the messenger should bear also a binary order). This might be a serious problem for a biological population (although it is not principally excluded: the operation of the immune networks involves tremendously large numbers [16] of different chemical molecules). In some cases, the amount of the required communication can be reduced by elimination of communication between a randomly chosen pairs of the population elements. Then the system dynamics is closer to that of the "diluted" neural networks where a large number of the synaptic connections is deleted. It is known that, at least in some problems of the information processing, the diluted networks can perform not much worse than the systems with the complete connectivity [17].

3. Mass Communication

In this section we examine the functional possibilities of an alternative mode of communication. In contrast to the above discussion, it is now assumed that the members of a population submit their messages into a common medium without addressing them. Thus, this class of models can be viewed as imitating, in a primitive form, the processes in the social systems which involve the "mass media".

The messages submitted to the means of mass communication have no addresses: they are intended to form a collective (or "public") opinion. A remarkably similar behaviour is observed in more simple systems, such as populations of biological cells or insect societies. The biological cells or individal insects release different chemical substances which spread diffusionally and mix in the medium. Since the rates of release of the mediators differ for different individuals and depend on their momentary states, the chemical composition in the medium reflects the current pattern of activity of the entire population and its role is analogous to the "public opinion" in the social systems.

The populations, which employ mass communication, can display very complex patterns of collective dynamics which are also functionally similar to the processes of information processing in neural networks.

One of the important functions realized by neural networks is the property of *associative memory*. The neural networks can keep in their memory a certain number of different "typical" patterns. When a new pattern is presented, they can retrieve in a response the nearest of the stored typical patterns.

This property can be achieved in a variety of ways, by choosing an appropriate synaptic matrix. The simplest of them was proposed in 1949 by Hebb [18] who suggested that the elements of the synaptic matrix should be constructed as

$$J_{ij} = (1/M) \sum_{\alpha=1}^{M} S_i^{(\alpha)} S_j^{(\alpha)} , \qquad (8)$$

where $S_i^{(\alpha)}$ is the state of neuron i in the stored typical pattern α; the total number of stored typical patterns is M. The dynamical properties of neural networks using the Hebbian matrix (8) were first analyzed by Hopfield [19].

Below we show (after [20,10,14]) how the behaviour of a neural network with a Hebbian synaptic matrix can be emulated by a population of elements which use only the mass mode of communication. For definiteness, the model is formulated in terms of a population of biological cells which can communicate by releasing and receiving the mediator molecules.

We assume that the momentary state of a cell i is described by a binary variable S_i. Our aim is to construct the communicating population of cells in such a way that it possesses a set of M different stationary patterns of activity. For each of these patterns α a certain state $S_i^{(\alpha)}$ of any cell i is specified. Evolution of the population from an arbitrary initial pattern should result in emergence of one of these stable "typical" patterns of activity.

The cells communicate by releasing a number of different mediators into the intercellular medium, which diffuse there and form a homogeneous distribution. Each kind of a mediator is associated with a particular "typical" pattern α.

We suppose that any cell j releases the mediator molecules of kind α only when its current state S_j coincides with the state $S_j^{(\alpha)}$ required by the respective prototype pattern α. The rate of release β is the same for all mediators. To prevent their accumulation, the mediator molecules are assumed to decay at a constatnt rate γ, the same for all kinds of mediators. Diffusion is fast enough and ensures permanent ideal mixing of reagents.

The kinetic equations for the concentrations m_α of the mediator molecules are

$$\dot{m}_\alpha = -\gamma m_\alpha + (\beta/2N) \sum_{j=1}^{N} (S_j^{(\alpha)} S_j + 1) . \qquad (9)$$

It can be easily checked that generation of the mediators, described by the last term in this equation, obeys the law which is formulated above.

We must also specify how the mediator molecules act on the cells. We will assume that each kind of mediator tends to establish that state of the cell i which is required by the respective prototype pattern. In other words, each mediator α transfers an *order* to all the cells to go into the states which correspond to the prototype pattern α. We see that, in contrast to our first model, the orders are not addressed: they are released into the common medium. However, each cell interpretes a received order in an individual way. It keeps in its memory how it should respond to the reception of such a signal.

Since, within a small time interval, a cell can receive many contradictory orders, a "voting" procedure is again employed, i.e. the actual next state S_i' of the cell is determined by the dominant order. Namely, we assume that

$$S_i' = \text{sign } h_i \qquad (10)$$

where

$$h_i = \sum_{\alpha=1}^{N} S_i^{(\alpha)} (m_\alpha - q) \tag{11}$$

and $q = \beta/2\gamma$ is a threshold concentration (if the concentration of the mediator α is smaller than this threshold, the cells tend to go into the state which is opposite to that implied by this mediator).

We consider the limit when the life-time of the mediators is much shorter than the intervals between the transitions in the cells. Under this condition the mediator concentrations m_α adiabatically adjust to the momentary states of the cells and equation (9) yields

$$m_\alpha = (q/N) \sum_{j=1}^{N} (S_j^{(\alpha)} S_j + 1) \quad . \tag{12}$$

Substituting these concentrations into (11), we obtain

$$h_i = (q/N) \sum_{j=1}^{N} \sum_{\alpha=1}^{M} S_i^{(\alpha)} S_j^{(\alpha)} S_j = q \sum_{j=1}^{N} J_{ij} S_j , \tag{13}$$

where J_{ij} is the Hebbian synaptic matrix defined in (8). The positive factor q that enters into (13) is not essential, since it does not change the *sign* of h_i which determines the next state of cell i.

Hence, we see that in the limit of short mediator life-times the above model of mass communication is effectively equivalent to a neural network with the Hebbian synaptic matrix (which is also known as the Hopfield model).

The analysis of the opposite limit with long life-times of the mediators was performed in [14]. It shows that the property of associative memory is retained in this case too. Remarkably, the concentrations m_α of mediators play in this model a role of order parameters: when a particular prototype pattern sets in the system, the concentration of the respective mediator becomes maximal.

Above we assumed that diffusion of the mediator molecules is so fast that ideal mixing is achieved and their spatial distribution is uniform. However, the spatial effects can be easily incorporated into the model (see [10]). When diffusion is not infinitely fast, the mediator produced by a given cell do not spread beyond a certain distance (which can be estimated as $(D/\gamma)^{1/2}$ where D is the diffusion constant and γ is the rate of decay). After this modification, a situation with a finite-range mass communication is thus described by the model.

We have formulated our model of mass communication in terms of the chemical cell-to-cell signalling. It is tempting, however, to discuss what might be its implications if we apply this mathematical model to a communicating human society. Then the role of the mediator molecules is played by the messages submitted to mass media and circulated there. These messages bear no addresses. Instead, each message α represents one of the possible M opinions on a certain issue. The distribution of the opinion frequencies m_α constitutes the pattern of a public opinion.

The issue under debate is which pattern of activity should be established in the society or, in other words, what should be the distribution of the social roles of its members. A given individual supports in this model all propositions that do not change its current, already established role, and votes agains a change in its status. While doing so, it keeps watching the reaction of other individuals. If the dominant order in the public opinion requires it to change its role, the individual accepts this (a model where a certain degree of "disobedience" is present can be also investigated, see [14]).

The evolution of a society that uses these rules of social interaction is very sensitive to its initial conditions. Depending on the initial distribution of social roles, the society evolves to one of its fixed different patterns. Taken collectively, it has then a property of "associative memory": when it starts from a certain initial activity pattern, it automatically chooses the closest of its steady prototype regimes.

4. Remote Sensing

Our last model [21] explores yet another possibility of communication. We consider below a population which employes only a single kind of a chemical mediator. The transferred information is now encoded in the *spatial distributions* of this mediator which can be sensed by the individual population members. To make our analysis more topical, we formulate this model using some properties of real biological neurons.

If we look at a real neural network of the brain, we would see that a synapse, forming a connection between two neurons, consists of two membranes separated by a very narrow gap. The mediator which is released from the presynaptic membrane diffuses through this gap and arrives at the receptors on the postsynaptic membrane. In this way the direct connection between the two neurons is realized.

However, it is often found [5,22] that a neuron which releases a mediator has no contacts with its neighbours. Instead the mediator is released into the intercellular medium. By the process of diffusion it can simulatneously reach a large number of receptors located on dendrites of other neural cells.

Thus, the neurons can interact as well in the indirect manner through the chemical medium into which they are immersed. Such interactions are especially important for the neuroendocrinic cells which, in addition to usual mediators, also release some neuroactive peptides and neuro-enzimes.

We see that the realistic neural systems of the brain differ significantly from the idealized networks of discrete units, i.e. of the formal neurons. The actual neurons are surrounded by the medium that contains various types of chemical agents secreted by the cells. The chemical composition of this medium and the spatial distribution of the mediators in it are constantly changing. These changes may influence the neural activity including the secretion process itself. Hence, besides of the direct connections through synapses, the neurons can also interact by emission and reception of different chemical signals.

If this view of a neural network is accepted, the central question is what might be be role played by chemical mediators in information processing. In the previous section we examined two different models of communicating populations which can be also used to describe the neural networks in absence of direct synaptic connections. In the first of these models the addressable mediators are employed which could act only on the neurons whose address coincides with the one carried by the mediator. In the second of our models each mediator was associated with a particular stored pattern. Below we consider a different model of a neural network without direct synaptic connections. Its principal difference consists in the fact that it uses a single sort of a mediator but still has the full potential to keep in its memory and to recall a large number of patterns. We formulate also a learning mechanism based on the processes of neural plasticity and effective chemotaxis.

In this last model it is assumed that the dendritic tree of any neuron is strongly branched and has a very large number of receptors. Then, in the simplest approximation, one can introduce the *local density* of receptors of a given neuron which is described by a continuous function of spatial coordinates. We assume also that the axons are strongly branched. For any neuron, we introduce the local continuous density $A(\mathbf{r})$ of axon terminals where the mediator is released. Thus, a neuron is pictured in this model by the *"clouds"* of its dendritic receptors and axonic terminals.

We suppose that the neurons (enumerated by index i) posses two kinds of dendritic receptors, activatory and inhibitory, for the same mediator. Hence, the receptive field of any neuron is described by two continuous distributions, i.e. by the density $R_{1,i}(\mathbf{r})$ of its activatory

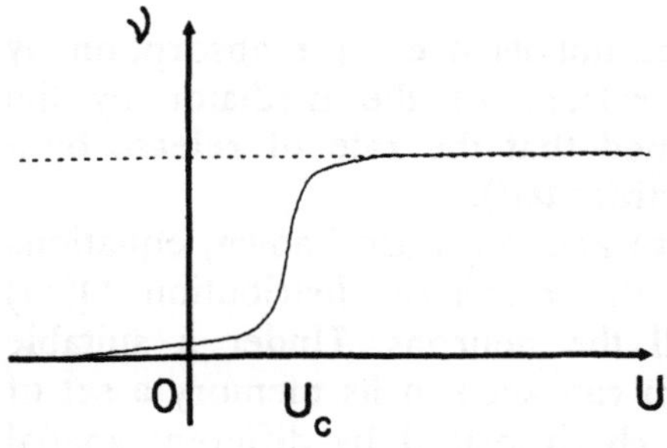

Fig.1. The typical form of the function $\nu[U]$.

receptors and by the density $R_{2,i}(r)$ of its inhibitory receptors. We denote the local concentration of the mediator by $Q(r,t)$.

In the presence of the mediator, an activatory receptor increases the electric potential U_i of the ith neuron by a contribution that is proportional to the mediator concentration Q at the location of this receptor. In the same case, the inhibitory receptor decreases the potential U_i by an amount proportional to Q. We neglect the delays caused by the finite speed of propagation of electrical signals inside neurons and assume that the momentary electric potential $U_i(t)$ of a neuron consists of a sum of simultaneous contributions from all its receptors, i.e.

$$U_i(t) = \int \alpha[R_{1,i}(r) - R_{2,i}(r)]\, Q(r,t)\, dr\,. \tag{14}$$

Here α is a positive coefficient.

We describe the dynamics of the neural activity in a simple way, assuming that the mean frequency ν_i of spikes generated by the ith neuron is a function of its potential:

$$\nu_i(t) = H[U_i(t)] \quad . \tag{15}$$

This function H[U] has a form of a smoothed step: the activity becomes significant only when a certain threshold U_c is exceeded (Fig.1). For large potentials U the function approaches a constant value. Below $\nu_i(t)$ is also called the *activity* of the ith neuron.

To complete the model, we write an equation for the mediator distribution:

$$\partial Q/\partial t = D\, \nabla^2 Q - \gamma Q + \beta \sum_i \nu_i(t) A_i(r)\,. \tag{16}$$

The first term in the right-hand side of (16) describes diffusion of the mediator, the second one takes into account its decay (or absorption by the cells). The third term describes release of the mediator by the neurons at their terminals. It is assumed that the rate of release by a neuron is proportional to its current activity $\nu_i(t)$.

When the distributions $R_{1,i}(\mathbf{r})$, $R_{2,i}(\mathbf{r})$ and $A_i(\mathbf{r})$ are known, equations (14)-(16) describe time evolution of the mediator distribution $Q(\mathbf{r},t)$ together with the activities $\nu_i(t)$ of all the neurons. Under a suitable choice of these distributions, the system can keep in its memory a set of prototype patterns. Such patterns are characterized by different spatial distributions $Q_i(\mathbf{r})$ of the mediator. Each prototype is associated with a certain neuron. When some initial distribution $Q_0(\mathbf{r})$ of the mediator is created, the system can detect its similarity to one of the stored prototype patterns and to retrieve the nearest prototype. At the same time the neuron, which is associated with the respective pattern, becomes activated.

It was shown in [21] that, in order to keep in the memory a set of N prototype patterns $Q_i(\mathbf{r})$, i = 1,...,N, the spatial distributions of the dendritic receptors and axonic terminals of ith neuron can be chosen as

$$R_i(\mathbf{r}) \sim \delta Q_i(\mathbf{r}) \quad , \quad A_i(\mathbf{r}) \sim Q_i(\mathbf{r}) \; . \tag{17}$$

Here $R_i(\mathbf{r}) = R_{1,i}(\mathbf{r}) - R_{2,i}(\mathbf{r})$ is the local difference in the densities of the activatory and the inhibitory receptors of the ith neuron and $\delta Q_i(\mathbf{r})$ is the local variance for the ith prototype pattern, defined as $\delta Q_i(\mathbf{r}) = Q_i(\mathbf{r}) - \langle Q_i \rangle$ where $\langle Q_i \rangle$ is the medium average for this pattern.

The underlying mechanism of this behaviour is simple: If the receptive field $R_i(\mathbf{r})$ of a certain neuron is closer approaching the presented pattern $Q_0(\mathbf{r})$, its initial activation is higher and, consequently, it begins to release the mediator at a higher rate. But, according to (17), the spatial distribution of its axonic terminals repeats, up to a constant component, the pattern of its receptive field. Therefore, the applied pattern evolves towards the prototype pattern kept by this neuron, until this pattern is established in the medium. When several neurons become initially active, competition between them starts. In its process, the neuron with the best initial fit has an advantage and eventually suppresses the activity of other neurons, i.e. the respective prototype mediator distribution $Q_i(\mathbf{r})$ emerges in the system.

Until now, we have not addressed the question of how the cell populations can *learn* the prototypes. The neurophysiological observations provide evidence [23] that learning may be related to neural plasticity, i.e. to the processes which involve growth (sprouting) of neurons, establishing new synaptic connections and modifying the strengthes of the already existing ones. The last two effects, however, cannot play a role in the considered model because the synaptic connections between the neurons are excluded.

It is known that sprouting of neurons is controlled by the trophic factors which represent some chemical substances. The best studied trophic factor is a protein, the nerve growth factor NGF. This protein stimulates the directed sprouting of axons and dendrites, by control of assembling the microtubules. It was demonstrated [24] that axons and dendrites adjust their directions of growth to the concentration gradient of NGF. According to a popular hypothesis, the peripheral target cells can produce NGF which directs sprouting of axons to their targets.

As already noted, our simple mathematical model is not intended to give a description of a real physiological system. Since we want only to demonstrate the possibility of such mechanism of learning, we assume below that the mediator itself is a substance which controls all plastic changes in the neurons and their sprouting. In our approach a neuron is modelled by a set of three continuous fields, specifying the local densities of dendritic receptors and axonic terminals. The individual receptors and terminals can be viewed as some particles which are distributed in the medium. Such particles can move, either together with a sprouting axon or dendrite to which they belong, or over the membrane of a resting neuron. Abstracting from the details of such motion, we can describe it phenomenologically as the "chemotaxis" of our particles, controlled by the mediator gradient.

We assume that the axon terminals tend to drift in the direction of the decrease of the mediator concentration. The drift occurs only when the neuron is active. The dendritic activatory receptors tend to drift in the direction of the increase of the mediator, if the neuron is active, and in the opposite direction if it is currently passive. This chemotaxial drift is imposed on random diffusional wandering of the particles (i.e. of the activatory receptors and the axonic terminals). The inhibitory dendritic receptors are uniformly distributed in the medium and their distribution is not changed during learning.

These processes are described by equations

$$\partial R_{1,i}/\partial t = D_1 \nabla^2 R_{1,i} - \mathrm{div}(p_1(\nu_i) R_{1,i}\, \mathrm{grad} Q) \quad , \tag{18}$$

$$\partial A_i/\partial t = D_a \nabla^2 A_i - \mathrm{div}(p_a(\nu_i) A_i\, \mathrm{grad} Q) \quad . \tag{19}$$

Here D_1 and D_a are the coefficients that characterize the rates of undirected diffusion-like spreading of the dendritic and axonic "clouds" in absence of the mediator gradient. Coefficients p_1 and p_a characterize the plastic sensitivity of dendrites and axons to mediator gradients. They depend on the momentary activity $\nu_i(t)$ of the considered ith neuron. In the simplest case such dependence can be chosen linear,

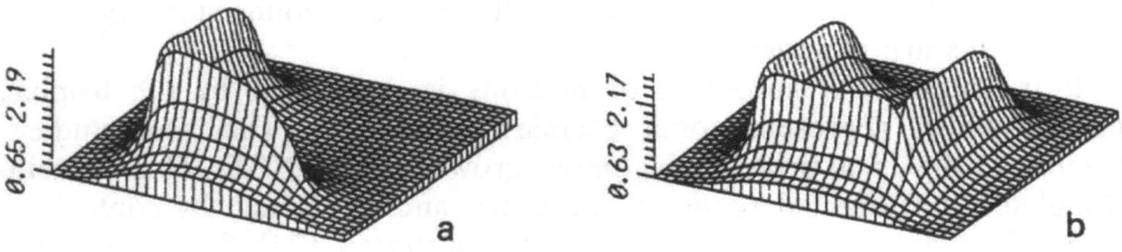

Fig.2. The axon fields of the first (a) and the second (b) neurons after learning.

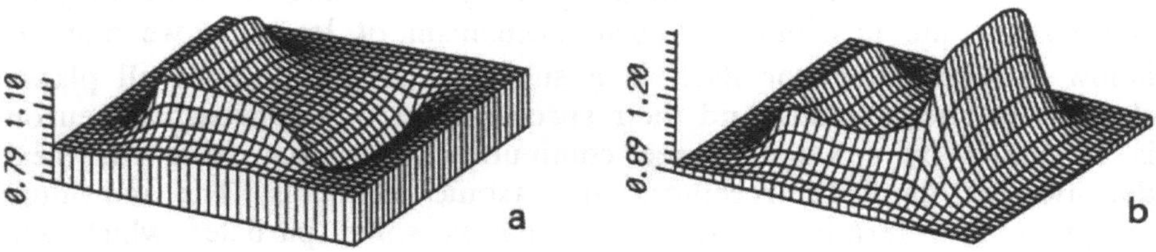

Fig.3. The activatory receptive fields of the first (a) and the second (b) neurons after learning.

$$p_1(\nu_i) = k_1(\nu_i - \theta) \quad , \tag{20}$$

$$p_a(\nu_i) = k_a\nu_i \quad . \tag{21}$$

According to (20), the activatory receptors drift in the direction of the mediator increase when activity ν_i of the ith neuron exceeds a threshold θ. When $\nu_i < \theta$, the direction of their drift is reversed. The axonic terminals always drift only in the direction of the gradient.

The learning procedure consists in the following. Suppose that we have a set of different distribution patterns of the mediator and want to teach the ith neuron to recognize only a particular pattern, characterized by the distribution $Q_i(r)$. Then we create this mediator distribution in the medium and keep it fixed. At the same time we keep the ith neuron in the active state (with $\nu_i > \theta$) and all other neurons in the passive states. When these conditions are maintained, the activatory receptors of the ith neuron drift towards the areas where the mediator concentration is maximal, while the activatory receptors of all other neurons tend to leave such areas. The axonic terminals of the ith neuron go into the regions with the higher mediator concentrations, whereas terminals of other neurons perform no directed drift.

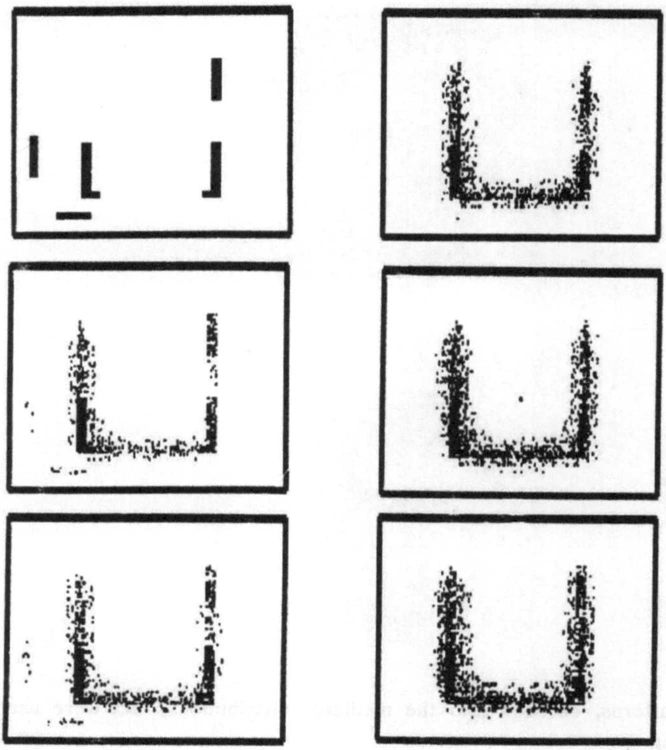

Fig.4. Reconstruction of the prototype pattern (the letter U) from the initial distorted image. The mediator distributions at equal time intervals are shown (top left to bottom right).

The training cycle consists of application in turn of all prototype patterns, each kept for a small time interval to produce changes in the dendritic and axonic fields. We expect that after many training cycles the system acquires the ability to recognize and reproduce the stored patterns.

To test the efficiency of this learning procedure, the numerical simulations were performed. In the first of them we trained the system, consisting of two neurons, to recognize the patterns which were the letters *L* and *U*. The details of the simulation are described in [21]. Fig.2 shows the axonic distributions $A_1(r)$ and $A_2(r)$ that were formed as a result of learning. We see that they follow the contours of the letters *L* and *U*. Fig.3 shows the final distributions $R_1(r)$ and $R_2(r)$ of the activatory dendritic receptors of both neurons.

Fig.5. Four prototype patterns, encoded into the mediator distributions, that were used in the process of learning.

The first of the learned patterns (letter *L*) differed from the second one (letter *U*) only in the absence of the right shoulder of the letter *U*. Therefore its presence in the pattern must have been an important discriminatory feature for the second neuron which was trained to recognize letter *U*. On the other hand, the important discriminatory feature for the first neuron, trained to recognize *L*, must have consisted in absence of this shoulder.

The receptive fields $R_1(\mathbf{r})$ and $R_2(\mathbf{r})$ that were obtained after training satisfied these requirements. We see that the receptive field of the second neuron (Fig.3b) has a strong maximum in the region where the right shoulder should be located. Contrary to this, the receptive field of the first neuron (Fig.3a) has a minimum in the same region. Both receptive fields have also (lower) maxima at the locations of the coinciding elements of the two prototype patterns.

After the learning procedure was finished, we used this system to recognize the prototype patterns *U* or *L* in the presented distorted images. Fig.4 shows the temporal evolution of the mediator distribution. The applied image, which was used to create the initial mediator

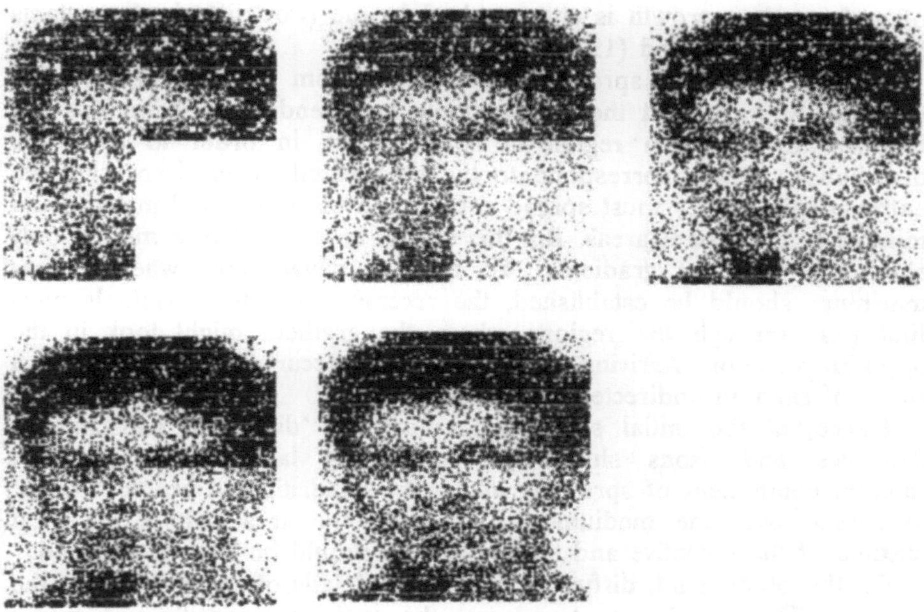

Fig.6. Reconstruction of the complete pattern under the system evolution. The mediator distributions at equal time intervals are shown.

distribution, represented a strongly distorted letter U (top of the left column). We can see how the missing elements appear and the superficial elements fade out in the process of time. The final mediator distribution (bottom of the right column) reconstructs the full prototype pattern.

The second computer experiment [21] differed in the number of patterns that were learned by the system. The set of the prototype patterns consisted of the digitized photographs of four different faces (Fig.5). Each of these patterns was associated with a particular neuron. After learning was finished, the system was presented with a test image. It was one of the four original photographs from which we had cut a quarter. The subsequent evolution of the system (Fig.6) resulted in the reconstruction of the complete pattern.

The learning procedure, which was employed in the above simulation, was slightly modified. Equations (18) and (19), which govern evolution of the dendritic and axonic distributions in the process of learning, include diffusion-like terms. These terms describe undirected sprouting of neurons, i.e a component of their growth which is completely random and does not follow the gradients of the mediator. The relative intensity of such random growth is characterized by the two "diffusion" constants D_1 and D_a in (18) and (19).

Note that random sprouting is essential from the point of view of learning. Suppose that the initial axonic and dendritic distributions are localized in a certain region of space. Then, in order to form the distributions which correspond to the delocalized patterns covering the entire medium, they must spread over it and establish local maxima and minima in the distant areas. But this cannot be a result of a motion only along the mediator gradients: To reach a distant area where a local maximum should be established, the receptors and the terminals must first pass through the regions where the gradient might look in the opposite direction. Arriving at such areas can occur only by chance, as a result of random undirected sprouting.

Hence, at the initial stage of learning the "diffusion" constants of dendrites and axons should be sufficiently large. The significant random component of sprouting allows the dendritic and axonic "clouds" to spread over the medium and to reach the areas where the local maxima of the receptive and terminal fields should be later established.

On the other hand, diffusion smears the details of the patterns. If the constants D_1 and D_a are too large, the evolving dendritic and axonic distributions cannot resolve the fine structure of the applied pattern, but catch only its rough features. Therefore, to ensure better resolution of the details, the smaller diffusion constants should be chosen.

To satisfy these two apparently contradictory conditions, the following procedure was employed. At the initial stage of learning, the higher values of D_1 and D_a were taken in the numerical simulation. They were then gradually diminished in the subsequent training cycles and thus the finer detailes of the digitized photographs were finally learned.

The above examples which involved recognition of visual images served only to demonstrate the possibility of associative memory and learning in the considered system. We have shown that a system of neural cells interacting only by release and absorption of a single mediator substance is already able to perform the complicated tasks of information processing. In a wider context, our analysis demonstrates that information can be effectively conveyed by spatial pattterns which are playing the role of the communication signals.

5. Conclusions

We have examined three typical examples of the communicating populations which employ different modes of communication between their members. The most complicated collective dynamics, which can emulate any process in a formal neural network, was found in the populations where the communication was based on the exchange of the addressed messages. However, this mode of communication may require too large amount of the information exhange. Although the potential collective behaviour of the populations employing mass communication is less complex, this mode of communication is still able to reproduce such important property of distributed information processing as associative memory. In the last of the considered examples we have explored the opportunities emerging in a population whose members are able to extract information from, and thus to produce a response to, the spatial distributions of the common mediator substance.

Our analysis was centered on clarifying the functional analogies between communicating populations and neural networks. Obvioisly, this represents only one side of the problem of collective dynamics in the communicating populations. To explain the processes of formation of complex spatio-temporal patterns in the populations of various biological species, more advanced mathematical models may be required, which would also take into account the motility of the species or assume a larger repertory of responses of the individual population members.

The incentive to studies of the population dynamics lies not only in an attempt to better understand the properties of real biological populations (which could open way to controlling their collective behaviour). The results of such studies can also be used in engineering (see [25]) of artificial "living" systems which are designed to execute complex technological tasks by mimicing the forms of collective behaviour of biological objects.

References

1. F. Alcantara, M. Monk "Signal propagation during aggregation in the slime mold *Dictyostelium discoideum*" J. Gen. Microbiol. **85**, 321-334 (1974)
2. G. Gerisch, D. Hulser, D. Malchow, U. Wick "Cell communication by periodic cyclic AMP pulses" Phil. Trans. R. Soc. Lond. B **272**, 181-192 (1975)
3. E. O. Budrene, H. C. Berg "Complex patterns formed by motile cells of *Escherichia coli*" Nature, **349**, 630-633 (1991)
4. J. D. Murray *Mathematical Biology* (Springer, Berlin 1989)
5. G. M. Shepherd *Neurobiology* (Oxford Univ. Press 1983)
6. A. W. Robards, W. J. Lucas "Plasmodesmata" Annu. Rev. Plant Physiol. Plant Mol. Biol. **41**, 369-419 (1990)
7. J. M. Pateels, J. L. Deneudourg (eds.) *From Individual to Collective Behaviour in Social Insects*" (Birkhauser, Basel 1987)

8. J. M. Tyson, K. Alexander, V. Manoranjan, J. Murray "Spiral waves of cyclic AMP in a model of slime mold aggregation" Physica D **34**, 193-207 (1989)
9. P. B. Monk, H. G. Othmer "Wave propagation in aggregation fields of the cellular slime mould *Dictyostelium discoideum*" Proc. R. Soc. Lond. B **240**, 555-589 (1990)
10. A. S. Mikhailov *Foundations of Synergetics I. Distributed Active Systems* (Springer, Berlin 1990)
11. *Parallel Distributed Processing*, Vol.1, eds. D. E. Rumelhart et al. (MIT Press, Cambridge, MA 1986)
12. H. Haken *Synergetic Computers and Cognition* (Springer, Berlin 1990)
13. A. S. Mikhailov, I. V. Mit'kov, N. A. Sveshnikov "Molecular Associative Memory" BioSystems **23**, 291-295 (1990)
14. A. S. Mikhailov "Information processing by systems with chemical communication", in *Rhythms in Physiological Systems*, eds. H. Haken, H. P. Koepchen (Springer, Berlin 1991) pp.339-350.
15. W. C. McCulloch, W. Pitts "A logical calculus of the ideas immanent in nervous activity" Bull. Math. Biophys. **5**, 115-137 (1943)
16. N. K. Jerne "Towards a network theory of the immune system" Ann. Immunol. (Inst. Pasteur) **125C**, 373-389 (1974)
17. B. Derrida, E. Gardner, A. Zippelius "An exactly solvable asymmetric neural network model" Europhys. Lett. **4**, 167- 171 (1987)
18. D. O. Hebb *The Organization of Behaviour* (Wiley, New York 1949)
19. J. J. Hopfield "Neural networks and physical systems with emergent collective computational abilities" Proc. Natl. Acad. Sci. USA **79**, 2554-2558 (1982)
20. A. S. Mikhailov, I. V. Mit'kov, N. A. Sveshnikov "Dual description and dynamics of the Hopfield model", preprint 89-50/127, Institute of Nuclear Physics, University of Moscow, 1989
21. E. M. Izhikevich, A. S. Mikhailov, N. A. Sveshnikov "Memory, learning and neuromediators" BioSystems **25**, 219-229 (1991)
22. K. Koketsu "Modulation of receptor sensitivity and action potentials by transmitters in vertebrate neurones" Jap. J. Physiol. **34**, 945-960 (1984)
23. M. Berry "Cellular differentiation: development of dendritic arborisation under normal and experimentally altered conditions" Neurosci. Res. Prog. Bull. **20**, 451-463 (1982)
24. F. Hucho *Neurochemistry* (VCH Verlagsgesellschaft 1986)
25. A. S. Mikhailov "Artificial life: an engineering perspective", in *Evolution of Dynamical Structures in Complex Systems*, eds. R. Friedrich, A. Wunderlin (Springer, Berlin 1992) pp. 301-312

From Social Engineering to Synergetics
On Metaphors, Models and Reality

Å. Andersson

Professor of Economics, Director of the Institute for Futures Studies, Stockholm, Sweden

Social engineering is a word growing in prominence since the turn of the last century. The concept has a metaphorical ring to it. One gets the impression that there is some economist or other social scientist working in a laboratory refining some tool or machinery to be used in the social system. But such a metaphoric interpretation is wrong. Until very recently the social and economic sciences and policies have had little or no experimental support.

The concept and the underlying theoretical ideas developed in the 19th century with the, by then, obvious possibilities of economic development and the equally obvious needs for social reforms. The French utopians were among the first to envisage a future where engineering principles would be brought into use in order to improve economic as well as social efficiency. Also the Fabians of early English socialism saw a similar possibility of rationalism in the formulation and execution of economic and social policy.

However, social engineering was not an idea of socialist thinkers only. In Scandinavia the foundation of a new school of social engineering was formulated by the liberal-conservative economist Gustav Cassel in his popular book *Socialpolitik*, published in 1902. In this book the rationalist and mathematically trained economist Cassel formulated a number of basic principles to be used in order to achieve socially sustainable economic growth. According to Cassel, no society can survive in prosperity unless it can create a growth inducing structure of social and economic relations. Each component of the large social machinery must have incentives to support economic progression. The book became an inspiration to many groups of social scientists (demographers, sociologists, political scientists, and economists). The economist Gunnar

Springer Series in Synergetics, Vol. 62 **Interdisciplinary Approaches to Nonlinear Complex Systems** – Eds.: H. Haken and A. Mikhailov

Myrdal, being the most famous student of Gustav Cassel, inherited Cassel's rationalist, engineering view of society. And Myrdal was to become a primary source of inspiration in the transformation of social engineering into a reasonably coherent theory of the welfare state.

Cassel was also probably the first theoretical economist to propose a balanced growth property of an economic system. According to this property the rate of balanced growth is determined by the ratio between the percentage share of the national income going into savings and the capital requirement per unit of national income. This balanced growth condition was later on to be refined by a number of economists, including Erik Lundberg (1937), Roy Harrod (1948) and Evsey D Domar (1951). It seems fairly obvious that Cassel was of the conviction that the labor market, the health care system and other social arrangements would have to be adjusted in such a way that the rate of savings could be kept at a high and stable level in order to insure sustained development. It is also obvious that social arrangements ensuring an efficient use of capital would be another means of generating a sustainable, high rate of growth of the economy, which according to Cassel was the only way of permanently securing an orderly society. It is no doubt that Cassel in his analysis of the means of social engineering was greatly inspired by German theorists of the 19th century and practices of social policy as implemented by Bismarck.

The Theory of Balanced Growth

Gustav Cassel's theory of equilibrium growth, as formulated in *Theoretische Sozialökonomie* (1917), was used as a starting point in a generalization of the theory by the mathematician and physicist John von Neumann in a paper published in 1936. Assuming a completely closed society with a large number of producers interconnected with each other by technological conditions of production, von Neumann proved that there would indeed exist a general equilibrium rate of growth and as a saddle point property a dual rate of interest that would ensure sustained economic growth with equilibrium proportions of all inputs and outputs of the expanding economy. For an exposition of the theory see e.g. Morishima (1964), Nikaido (1968), Zhang (1991) or Andersson (1968).

The Cassel property of balanced growth can be proved with the aid of a slightly simplified version of the von Neumann theory. In this model of a growing economy we make the simplification that the economy can be subdivided into a finite number of sectors, each sector producing one commodity only. Thus, the index of a sector is also an index of a commodity. We also make the simplification that there is one recipe only for the production of a given commodity by a sector. This implies that there is no possibility of substituting different inputs in the production of an output. As in the von Neumann model we assume that all households can be aggregated into a sector producing labor by inputs of given amounts of products per unit of labor delivered. These simplifications were proposed by Leontief (1953). We also assume that the economy can be closer to or further from full use of capacity, as indicated by μ. A sustainable equilibrium is such that there is full use of capacity and a rate of growth of capacity compatible with this level of capacity use. In order to show the dual properties we solve this problem by maximizing the rate of capacity use at some given, uniform, rate of growth of capital and production.

$$\max_{\{x,\ \mu\}} \mu$$

subject to

$$\mu\, x_i = \sum_j a_{ij}\, x_j + \lambda_w \sum_j b_{ij}\, x_j$$

where

μ = rate of production capacity utilization

x_i = production of good i

a_{ij} = use of current input i per unit of output of j

λ_w = warranted rate of growth (given)

b_{ij} = capital input i per unit output j

This optimization problem corresponds to

$$\max_{(x,\, \mu,\, p)} L = \mu - \sum_i p_i \left(\mu\, x_i - \sum_j a_{ij}\, x_j - \lambda_w \sum_j b_{ij}\, x_j\right);$$

with the necessary conditions of a maximum:

$$\frac{\delta L}{\delta \mu} = 1 - \sum_i p_i\, x_i = 0$$

$$\frac{\delta L}{\delta x_i} = -\left(p_i\, \mu - \sum_k p_k\, a_{ki} - \lambda_w \sum_k p_k\, b_{ki}\right) = 0$$

$$\frac{\delta L}{\delta p_i} = -\left(\mu\, x_i - \sum_j a_{ij}\, x_j - \lambda_w \sum_j b_{ij}\, x_j\right) = 0$$

We now assume a rate of growth $\lambda^* = \lambda_w$ sufficient to achieve full use of production capacity, i.e. such that $\mu = 1$.

We then have the balanced growth condition

$$\lambda^* = \frac{1-\sum_k (p_k/p_i)\, a_{ki}}{\sum_k (p_k/p_i)\, b_{ki}} = \frac{\textit{profit ratio}}{\textit{capital-output-ratio}} = \frac{\textit{real gross savings ratio}}{\textit{capital-output-ratio}}$$

in <u>value</u> terms (i = 1, ..., n).

Increasing profits or a decreasing capital-output ratio will thus imply an increasing balanced rate of growth.

Inflation, i.e. increasing all prices by the same percentage, will leave the growth rate unchanged, provided the balanced rate of growth equals the real rate of interest, which is defined as the difference between the nominal rate of interest and the rate of inflation of price.

Optimal Control of the Economy

In the same year as the publication of Von Neumann's path-breaking article on conditions of balanced growth a counterpunctual publication was released. This publication was *The General Theory of Employment, Interest and Money* (1936) by John Maynard Keynes. Keynes essentially made the proposition that we should not expect balanced growth but rather permanent stagnation at a depressed level of the economy unless the state would intervene directly in the investment process. Because of the importance of time, as manifested by accumulated production capacity and expectations, any equilibrium would be brittle, according to Keynes. Any downward turn of profitability would easily be transmitted into expectations of further profitability problems, causing dramatic downturns in the willingness to invest. By positive feedbacks investments, production and employment would dwindle into lower levels than full capacity and finally stagnate at an under-employment equilibrium.

The focus of *General Theory* was on methods of achieving full employment by the use of different instrumental variables or controls, available to the government. It was then rather obvious that public investments would be a suitable instrument, although it was also clear that a lowering of the rate of taxation would be an alternative way of increasing demand, production and thus employment.

Keynes seminal contribution triggered an almost immediate respons, primarily in Scandinavia and the Netherlands and to some extent also in the USA. The Keynesian analysis seemed to be completely in line with the ideas of proponents of social engineering and other forms of government intervention. Three economists ought to be mentioned in this connection: Ragnar Frisch (Norway), Jan Tinbergen (the Netherlands), and Bertil Ohlin (Sweden). Frisch and Tinbergen were to receive two of the first Nobel Prizes in economics.

Bertil Ohlin basically agreed with Keynes in favouring state intervention to achieve full employment, although Ohlin was rather skeptical to the consequences in terms of under-balanced government budgets. Later on, Ohlin became one of the proponents of interventions in the market economic system by the use of "constraint planning". Frisch and Tinbergen represented a much more forceful view on the role of the government in guiding the market economy. They were firm believers in mathematical economics and the branch of statistics called *Econometrics*. According to their view, economics ought to be reoriented into a primarily quantitative science, with the research program closely resembling that of theoretical physics. This would not only assure a high degree of consistency in studying interdependency within the economic system. The formulation of quantitative economic relations would also simplify estimation and simulation of quantitative economic processes.

Already by 1939, Tinbergen had formulated and estimated a quantitative version of the core model of *General Theory*. This attempt meet with immediate disapproval by Keynes in a review article in the *Economic Journal*. This was to be expected. Keynes had already in *General Theory* formulated his views about the risks of mathematical modelling in economics:

"It is a great fault of symbolic pseudo-mathematical methods of formalising a system of economic analysis, such as we shall set down in section VI of this chapter, that they expressly assume strict independence between the factors involved and lose all their cogency and authority if this hypothesis is disallowed; whereas, in ordinary discourse, where we are not blindly manipulating but know all the time what we are doing and what the words mean, we can keep "at the back of our heads" the necessary reserves and qualifications and the adjustments which we shall have to make later on, in a way in which we cannot keep complicated partial differentials "at the back" of several pages of algebra which assume that they all vanish. Too large a proportion of recent "mathematical" economics are mere concoctions, as imprecise as the initial assumptions they rest on, which allow the author to lose sight of the complexities and interdependencies of the real world in a maze of pretentious and unhelpful symbols."

This view on the possible misuses of mathematics in economics were not the words of a pure layman. Keynes had a fairly sophisticated training in mathematics and was regularly involved in philosophical and other discussions about the role of mathematics in probability theory and economics with B Russel, F P Ramsey and other philosophers and mathematicians.

Social engineering and interventionist modelling

Keynes had opened a gate that he could not close. In the 1940's, 1950's and 1960's, a steadily increasing flow of papers and books, oriented to finding everything from theorems to rules of thumb of quantitative economic policy making, appeared in the steadily increasing number of journals devoted to such theories and models. In economics Dutch, Scandinavian and increasingly American scholars contributed to this development. Almost simultaneously, Tinbergen (1956) and Bent Hansen (1955) formulated a static version of a control theoretic rule of macro economic policy making. The rule was based on the following theoretical argument.

Assume the existence of a set of implicit functions f_i relating a finite set of economic and other variables to each other.

$$f_i\left(x_1, ..., x_k ; x_{k+1}, ..., x_n\right) = 0;$$

$$(i = k + 1, ..., n).$$

Assume further that variables $\{x_1, ..., x_k\}$ are considered of importance to the well-being of the citizens of the economy. These variables are then seen as targets or goals of economic policy. If these targets are to be reached, there must be sufficiently many instruments available to reach these targets. The instruments must then be regarded as variables to be freely adjusted so as to reach the prescribed (fixed) target levels. A consistency of goal formulation would require that the number of instrumental variables is at least as large as the number or prescribed policy targets. This rule of rational economic policy was proved more or less reasonably by mathematical methods, but the motivation was often in terms of metaphoric arguments. It can e.g. be argued that any hunter aspiring to kill n animals (targets) would be wise to bring at least n bullets (instruments) to the hunt.

One of Tinbergen's most prominent student, Henri Theil (1963) refined the *ends-means*, *targets-instruments* modelling of interventionist policies by dissolving the distinction between target and instrument variables. According to Theil there is no possibility to make such a distinction. Most voters and other participants in the political processes would not be able to make any clear statements about such a status of some macro-economic variable. One example would be the tax rate. To Tinbergen, Hansen and most of the other interventionist economists the tax rate would be an instrumental variable, while Theil would argue that this would be grossly at variance with the common views of politicians, voters and other citizens. The distinction would in general be blurred. But according to Theil this would not be any essential reason for abandoning quantitative interventionism or social engineering practices. In his *Optimal Decision Rules for Government and Industry,* he proposed a model of optimal compromises in economic policy. By measuring some norm of deviations from the ideal values of instrumental and goal variables an optimum compromise could be found, even in situations which would be deemed inconsistent according to the classical rules of economic policy, as formulated by Tinbergen and Hansen. In order to clarify the structure of the Theilian proposal for modelling of optimal interventions we use the following simplified model. The instantaneous rate of growth, y, the instantaneous rate of inflation, p, the rate of public capital accumulation, I, and the rate of interest, r, are assumed to be related to each other by some implicit function $T(y, p, r, I) = 0$.

The rate of public capital accumulation, or public investment, and the rate of interest are assumed to be instruments of economic policy but only in the sense that the government can freely determine their values by decisions. The government would like to keep the four variables of the problem at some ideal level, indicated by an *. Any deviation from these ideal values would imply a loss of welfare to the population (or at least to the government). Theil then proposes the use of squared deviations from the ideals as measures of these welfare losses. Thus, an optimal decision rule for government would imply solving the following contrained minimization model.

Minimize $W = 0.5\,(\omega_i\,(y - y^*)^2 + \omega_p\,(p - p^*)^2 + \omega_I\,(I - I^*)^2 + \omega_r\,(r - r^*)^2)$

Subject to $T(y, p, I, r) = 0$;

where T ( ) is a concave transformation function connecting the target and instrument variables to each other. The T-function is assumed to be r-differentiable. Five marginal conditions of an optimum policy can be derived:

$$\omega_i \left(y-y^*\right) - \lambda \frac{\delta T}{\delta y} = 0 ;$$

$$\omega_p \left(p-p_*\right) - \lambda \frac{\delta T}{\delta p} = 0 ;$$

$$\omega_I \left(I-I^*\right) - \lambda \frac{\delta T}{\delta I} = 0 ;$$

$$\omega_r \left(r-r^*\right) - \lambda \frac{\delta T}{\delta r} = 0 ;$$

$$T \left(y, p, I, r\right) = 0 .$$

These conditions can be summarized as the following three policy rules:

$$\left(I-I^*\right) = \left(\omega_y \,/\, \omega_I\right) \left(\frac{\delta T}{\delta I} \Big/ \frac{\delta T}{\delta y}\right) \left(y - y^*\right)$$

$$\left(r - r^*\right) = \left(\omega_p / \omega_r\right) \left(\frac{\delta T}{\delta r} \Big/ \frac{\delta T}{\delta p}\right) \left(p - p^*\right)$$

$$T \left(y, p, I, r\right) = 0$$

These optimal decision rules state that the public investment should be determined so as to be at a deviation from the ideal level that would be marginally proportional to the deviation of the growth rate from the ideal. The deviation of the rate of interest from the ideal level should be similarily marginally proportional to the ideal rate of inflation. If the transformation function would be linear, then the optimal decision rule would be a standard linear feedback.

Theil and especially many of his followers enlarged this approach to increasingly complicated situations of government intervention. Dynamic and stochastic variants were proposed to be used in connection with econometric models of increasing size in terms of the number of variables and interdependencies between variables.

Meanwhile, other large scale models for social engineering had appeared in seemingly unrelated fields. The planning of land use and transportation was increasingly modelled by large scale simulation or optimization models explicitly intended to provide guidance for policy makers. One prominent class of such models was based on Lowry (1963). Others were based on linear and nonlinear programming methods as proposed by Dantzig (1954), Koopmans (1965) and others. In all these logically consistent models it was assumed that reality could be smoothed into a linear or at least convex structure. If such assumptions could be made, optimal interventions by social engineering would be ensured not only at the level of the macro economy but also at the more detailed levels of land use and transport flows optimization. In a sense the models were built on an assumption that society could be seen as a mechanical machinery composed of smaller component mechanical machines. This turned out to be an assumption that could only be upheld under very special social, political and economic circumstances.

Interaction and loss of predictability

Increasing realism in modelling became the threat to modelling by social engineering methodology. One of the basic characteristics of economic theory is the stress on interdependencies between decision makers. And interdependency inevitably generates model complexity. This was probably the reason why Keynes resisted the attempts to transform *General Theory* into a set of linearized and essentially static equations. Especially investments are regulated in some complex interactive patterns. These patterns cannot be reduced away by linearization. Puu (1992) has shown that an even mildly nonlinear, interactive, investment response in a dynamic model of an economy would generate an unpredictable or chaotic motion of national income and investments. Similarly, models of interactions in oligopolistic markets tend to be characterized by non-optimal, excessively stable solutions or chaotic fluctuations of prices and quantities.

Economic and social interaction breeds unpredictability. The empirical volatility of the markets for energy and currencies during the last two decades has meanwhile discouraged most social engineers, planners and other believers from quantitative modelling of large systems.

Synergetics - A Way Out of Chaos in Modelling Economic Development?

The formulation of econometric and other mathematical models, based on the social engineering and interventionist modelling strategy, were mostly static, if nonlinear, or linear, if dynamic. When used in planning, the nonlinear and static models were mostly used to generate a scenario of some future combination of means and ends. The inherent difficulty of such a scenario technique is to find a consistent trajectory (or traverse) between the initial state and the final state, according to the scenario. In most cases it cannot even be shown that a viable trajectory between two, sufficiently distant states could be found, even in principle.

The problem with the linear or linearized dynamic models is even worse. A linearization is almost always permissible only within a short span of time (i.e. for a specified short interval of the range of the different variables). Using a linearized version of an inherently interdependent and nonlinear model generically leads to infinite exponential growth, decay or a fixed point solution, which are model results grossly at variance with real tendencies of economies and social systems. Experiences of simulation with large, nonlinear economic and social models have rarely been successful. The reason is by now clear and obvious. Basically static reasoning has been mechanically transformed into systems of nonlinear difference or differential equation systems, without properly analysing the **character** of the variables and processes in terms of dynamics and scope of impact upon the other variables of the system. Undifferentiated time-scales and symmetric treatment of all variables and their interactions with each other is no way of avoiding chaos in a complex dynamic model. As shown by Haken (1983) a careful analysis of the real world in these respects is needed to achieve an observable, predictable and controllable dynamic system that would otherwise be hard to observe, of limited predictability and totally lacking controllability.

Economic and social variables can be decomposed into two basic types. The first type is characterized by **privateness** in their consequences. Such a private good will have consequences for an individual household or firm only. Other goods are **public** in the sense that the availability of the good will have an impact on the level of productivity or utility of many firms or households. It must be stressed that a public good can have a differential impact upon different households or firms. E.g. one firm, producing a chemically highly sensitive product can be greatly supported by the improved availability of clean air, while another firm would have a limited advantage of the same public good. It should also be remarked that there are also public "bads", e.g. pollution, influencing many firms and households, simultaneously.

A few economic variables are highly public. One example is the construction of property rights, which will influence all firms and households of a given economic and political region. Public goods are in important senses similar to order parameters of the natural sciences. However, public goods are often hard to handle because of the fact that social and economic behaviour is determined by intentional decisions.

Goods and economic processes can also be decomposed according to dynamic characteristics. Some goods (or phenomena similar go goods) are of great durability as economic goods. A building generates a stream of services during a lifetime of decades or centuries, while milk would be destroyed within a few days. Similarly, theorems or laws of nature would have an almost infinite durability, while most information relayed by the media would lose any impact within a few hours.

A few goods and phenomena are public in their consequences and inherently slow in change processes. Characteristics such as average speed of communication and transportation on networks are such public goods, normally changing very slowly over time and only by slow and steady improvements of many links and saddle point characteristics of a network. Similarly the efficiency of mathematical algorithms and other solution methods are slow and public. Networks and knowledge, including values and decision rules, are simultaneously slow and public phenomena. These goods are in the old economic literature often called **infrastructure**. Much of this infrastructure can be represented by slowly changing order variables as exemplified above. When modelling a dynamic process predictability and controllability can be improved substantially if a proper subdivision is made between the slow and public infrastructure variables and the fast or private goods.

The use of these procedures can be illustrated with the following system of differential equations:

$$\dot{x} = T \cdot \omega_i (x, k, c) \cdot x_i^{\alpha} \cdot (\bar{x}_i - x_i)^{\beta} \; ; (i = 1, ..., n)$$

$$\dot{k} = T^{-\lambda} \cdot F (k, c, x) \; ; T, n, \lambda = \textit{positive integers} \; ;$$

$$\dot{c} = T^{-n \cdot \lambda} \cdot G (c, x) \; ;$$

where

x_i = employment in private sector i;

$\bar{x}_i$ = externally determined maximum value of x_i;

ω_i = marginal productivity of labor;

k = stock of knowledge;

c = communication network capacity;

k, c together constitute the infrastructure of this dynamic economic system provided that T, n or λ is sufficiently large, there is a possibility of an adiabatic approximation of this dynamic economic system.

Marginal productivity of labor is normally possible to approximate with a polynomial function of the level of employment. After a certain critical level of employment the marginal productivity declines with increasing employment. Larger availability of knowledge and network infrastructure is, according to most empirical studies, factors that would increase marginal productivity of labor of all or most firms, simultaneously. If there is a slow growth of infrastructure the result will sooner or later be a drastic transformation of the employment structure according to this system of differential equations. However, most of the time there will an equilibrium solution, determined by the dynamics of competition between employers, where the equilibrium is determined by equalization of marginal productivities (and wage rates). The essential difference between infrastructure and employment of this model can be seen metaphorically as a relation between an arena or stage and the **games** played in a labor market, subject to the constraints given by the imperceptibly slow changes of the arena.

With such a synergetic recontruction of economic theory, there is also an emerging new view of social engineering or economic policy interventions in the economic development process. It should be obvious from this analyses that there is little room for classical social engineering based on difference or differential equations moving on an undifferentiated time-scale, while all variables are treated equally. Rather, our reconstruction clarifies the division of labor between the market and politics. Political interventions directly in the market are of little value according to this view of the matter. The political system should rather be concentrated on the slow and public variables or the infrastructure determining the qualitative patterns of the markets in the long run, thus achieving not only predictability but also possibilities of ensuring sustainable development.

Infrastructure and Sustainable Development: On Complexity and Sustainability

Knowledge belongs to the infrastructural arena determining the structural outcome of economic and ecological development. Unfortunately, knowledge is a slippery concept, when applied to the theory of economic dynamics. Often the stock of knowledge is simply measured as the total number of school years accumulated by the population.

This is at best an unsophisticated proxy for things that ought to be measured. The problem with this measure is that it puts the economy of the former Soviet Union at a parity with economies that are obviously much more developed in terms of technologies of production and quality of products.

In this final section of the paper there is no intention to present any complete solution to this problem but rather to indicate a reconceptualization that would make the synergetic interaction between knowledge, networks and economic and ecologic development more obvious than with conventional approaches of sustainable development theory.

A candidate as a basic concept in this reconceptualization is **product and process complexity.** By complexity of the product I mean the minimal length of the decsription of the characteristics and properties fully representing the product. A string quartet by Beethoven is this sense a much more complex product than any childrens canon, which is basically an iterative procedure easily described to any child of some musical talent.

Process complexity can similarly be defined as the minimal length of the recipe needed to completely and accurately describe the procedures involved in producing some good. (This definition is rather close to the computer program complexity definitions suggested by Chaitin (1966) and Kolmogorov (1968), see also Cover (1974)).

The essential characteristic of product and process complexity is the need for a knowledge base in order to generate and apply complexity to the production system. Complexity in the sense defined above has a unique inormational content but is obviously contextually sensitive. The minimality of the length of an instruction cannot be determined unless the educational and communication capacities are predetermined. It is impossible to use some process instructions in a low education society, if these instructions have been generated in an environment of high average education. Similarly, a product of high complexity cannot be used by people who have not achieved the necessary level of knowledge needed in the use of such a products. A Stradivarius violin is an excellent example of both kinds of contextual interdependency of complexity.

Tentatively, the use of this conceptualization can be illustrated by the following diagram.

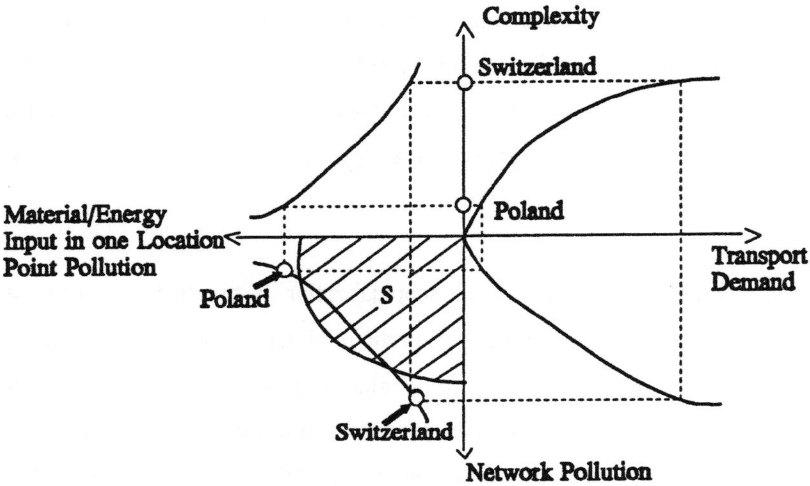

S = Sustainability Set

In the diagram we have assumed that the value of a commodity is determined by the complexity of the product. The complexity of the product is assumed to have been determined by the proper use of the stocks of knowledge, available. The level of product complexity is furthermore assumed to influence two variables. On the one hand, increasing the level of complexity (as the economy wide average) means a substitution of energy and materials for a more sophisticated product structure. This implies that the increasing complexity of products and processes will reduce the inputs of energy and materials per unit of value of the output. Secondly, increasing complexity requires (in the normal case) a search for more sophisticated inputs, requiring larger amounts of transportation inputs. A society of low product complexity (e.g. Poland) would thus have a combination of large uses of energy and raw materials at concentrated factory locations, while a country of high average complexity of products and processes (e.g. Switzerland) would be characterized by a combination of diffuse pollution by transports but very small local pollution levels by use of energy and raw materials. As indicated by the diagram both countries could be outside of the sustainability set, unless the infrastructure had been constructed in such a way so as to accomodate such a combination of complexity and ecological impacts.

Conclusion

This paper is intended to indicate certain methodological and conceptual problems associated with traditional economical theory, when applied to the dynamic development of an economy and the associated ecological system.

It is argued that synergetic approaches are of great use in the reconstruction of the theory of economic dynamics. Especially, the possibilities of subdivision of goods, according to their publicness and dynamic processes, according to their typical speed of change turns out to be of great value in improving predictability and controllability of complex economic systems.

Finally, a reconceptualization in the treatment of knowledge and its use in the production system is suggested. Two new concepts - product and process complexity - are suggested. The usefulness of these new concepts is illustrated in a heuristic model of ecological and economic interactions during a transformation from a low into a high knowledge society.

References

Andersson, Å.E. (1968) "From Interest and Prices to Capital and Growth". *Swedish Journal of Economics*, No. 4, 221-241.

Cassel, G. (1902) *Socialpolitik*. Gebers förlag, Stockholm.

Cassel, G. (1917) *Theoretische Sozialökonomie*. Stockholm.

Chaitin, G. (1966) "On the length of programs for computing finite binary sequences". *J. ACM*, vol. 13, October 1966.

Cover, T.M. (1974) *Universal Gambling Schemes and the Complexity Measures of Kolmogorov and Chaitin*. Technical Report No. 12, Statistics Department, Stanford University.

Dantzig, T. (1954) *Number, the Language of Science*. McGraw-Hill, New York.

Domar, E.D. (1951) *Essays in the Theory of Economic Growth*. Oxford University Press, New York.
Frisch, R. (1933) "Propagation, Problems and Impulse Problems in Dynamic Economics". *Economic Essays* in Honour of Gustav Cassel. London.
Haken, H. (1983) *Advanced Synergetics*. Springer-Verlag, Heidelberg.
Hansen, B. (1955) *Finanspolitikens Ekonomiska Teori*. Almqvist & Wiksell, Stockholm.
Harrod, R. (1948) *Toward a Dynamics Economics*. MacMillan, London.
Keynes, J.M. (1936) *The General Theory of Employment, Interest and Money*. Macmillan & Co. Ltd, London.
Kolmogorov, A.N. (1968) "Logical basis for information theory and probability theory", *IEEE Trans. Inform. Theory*, vol. IT-14.
Koopmans, T. (1965) "On the Concept of optimal Growth" in *The Econometric Approach to Development Planning*. Rand McNally, Chicago.
Lowry, I. (1963) *A Model of a Metropolis*. Rand Corporation.
Leontief, W.W. (1953) *Studies in the Structure of American Economy*. Oxford University Press, New York.
Lundberg, E. (1937) *Studies in the Theory of Economic Expansion*. Reprinted by Kelley & Millman, New York (1955).
von Neumann, J. (1936) "A Model of General Economic Equilibrium". *Review of Economic Studies*, 33, 1-9. (English translation from German original).
Morishima, M. (1964) *Equilibrium Stability and Growth - A Multi-sectoral Analysis*. Oxford University Press, New York.
Nikaido, H. (1968) *Convex Structures and Economic Theory*. Academic, New York.
Ohlin, B. (1934) *Penningstatistik, offentliga arbeten, subventioner och tullar som medel mot arbetslöshet: Bidrag till expansionens teori*. Norstedts, Stockholm.
Puu, T. (1992) "A Chaotic Process with Slow Feedback: The case of business cycles" in *Economic Evolution and Demographic Change*, eds Haag et al. Springer-Verlag, Heidelberg.
Theil, H. (1963) *Optimal Decision Rules for Government and Industry*. North-Holland, Amsterdam.
Tinbergen, J. (1956) *Economic Policy; Principles and Design*. North-Holland, Amsterdam.
Zhang, W-B. (1991) *Synergetic Economics*. Springer-Verlag, Heidelberg.

Social Order
From Individual Activity to Functional Cooperation

G. Küppers
University of Bielefeld, Fed. Rep. of Germany

To demonstrate that functionality and goal-directedness could be the outcome of blind interaction has proved to be a tricky problem for modern science. Its program of tracing all observable changes in nature back to the interaction of matter reduces Aristotle's multidimensional schema of causality to the *causa efficiens*. This "effective cause" was only one and, indeed, the least important of the four components in Aristotle's concept of causality. The other three were causa formalis, causa materialis, and causa finalis. It has often been noted that this explanatory schema is borrowed from a (manual) production process. A house is constructed by imagining a purpose, designing a structure, supplying certain building materials, and finally starting to build. The modern concept of force retains only the latter: A force is the cause of a change in a material substrate. With the omission of form-giving causality, there was no longer any plausible explanation within science for the way things are formed; and with the loss of purpose-giving causality, there was no longer any plausible explanation for the organization between things. The problem of establishing order, which the Greeks only needed to explain as far as the creation of the world from disordered chaos was concerned[1], and which Aristotle resolved with his theory of a world that had always been formed and was therefore eternal, now became a problem that pervaded all segments of reality. How can order, be it externally organized or self-organized, in any way arise without organizing powers? How can order arise or be maintained from forces that cause nothing other than changes of movement while remaining "blind" with regard to goals?

A solution to this problem was only found during the middle of the present century, when several branches of science independently discovered principles that were able to explain the origins of order and complex organization out of "blind" material interaction under labels such as synergetics, self-organization, dissipative structures, autopoiesis, and so forth.[2] Nowadays, the self-organization of matter is, in many cases, a well understood phenomenon in all domains of nature. The self-organization of social systems has also been a topic of self-organization research for many years. Interest here focuses on two issues: The first addresses the (self-) organization of patterns of collective behavior in a group of relatively homogeneous individuals whose changes in behavior depend on the behavior of the other members of the group. The social "interaction" here is restricted to the impact of the behavior of others on potential behavior modifications in an individual. These analyses are oriented toward the concept of changes in physical states, for example changes in magnetiza-

[1] As in Timaios and in the mythological forerunners.

[2] On the history of self-organization research, compare Krohn, Küppers, and Paslack (1987, pp. 441-465).

tion in a ferromagnet[3] and social activities as forms of *social interaction* (cooperation and communication) are disregarded.

The self-organization of social interaction is the focus of modern systems theory in sociology, its most prominent representative being Niklas Luhmann.[4] Here, interest concentrates on the development of special functional systems (economy, politics, science, etc.) within a concept of society as an autopoietic system in which communication is not only the element but also the basic form of social interaction. Society is communication and, as such, a network that continuously regenerates its elements (communication) precisely through this network. However, the differentiation of specific functional (sub-)systems within this approach is not performed by mechanisms of self-organization but through analytically introduced categories for describing model systems. Luhmann's binary schematizations - for example, true/false for the science system - are theoretical categories and not mechanisms that generate differentiations. Therefore, they are unsuitable for analyzing processes of system differentiation and system dynamics in real social systems.

In contrast, the approach to the self-organization of social systems presented here pursues a strategy of managing without any such problematic a priori decisions. It attempts to identify the mechanisms of self-organization that are available to the system for the formation of boundaries and system differentiation and that subsequently determine the system dynamics.[5]

How far theories of self-organization from the natural sciences can be generalized to social systems will depend on whether it is possible to generalize the mechanisms of natural self-organization to social processes without trivializing them in an unacceptable way. To test this issue, I will initially discuss the mechanisms of self-organization in natural systems. Then I will try to identify corresponding mechanisms in the social world and name the conditions of social self-organization. Finally, I will present the social system of science as an example on which we will demonstrate the successful application of theories of self-organization from the natural sciences.

1. Mechanisms of Self-Organization

A system, that is, an entity that can be differentiated from its environment is self-organizing (a) when it is autonomous, that is, when all changes in its state[6] are a consequence only of its internal operations (no external forces), and (b) when this autonomy is maintained by the system. A self-organizing system is accordingly a closed network of operations whose autonomy is maintained by this network. Autonomy does not mean isolation from the environment. However, environmental influences can only trigger changes in the system but

[3] Compare Weise (1990).

[4] Compare Luhmann (1984).

[5] Nobody would want to dispute that, in real societies, complex forms of social interaction exist on all levels and across the boundaries of functional subsystems. Admittedly, it remains unclear how far these social structures are an outcome of an (internal) self-ordering of social interactions or have other, external causes. Without doubt, external conditions can lead to the formation of social order; however, the decisive question is whether they are the cause or merely the trigger of order.

[6] We define state as the values of all relevant system variables at a specific point in time.

cannot cause them. Independent from autonomy, the system must be open for energy, matter, and information from the environment. Self-organizing systems are not in a thermodynamic equilibrium. An autonomous (closed) network of operations can only arise if the output of each single operation becomes the input of another operation: Cause and effect must be mutually dependent. Kant had already formulated this condition for self-organization in 1775, when he set himself the task of reducing not just the order of planetary orbits to mechanical principles like Newton but also their origins.[7] Examples of such circular, closed cause-effect chains can be found everywhere. In the laser, for example, laser light and the elementary emission processes of laser atoms are mutually dependent; in cellular convection, a specific temperature profile creates a flow within the liquid that, in turn, changes the temperature profile. In plasmas, for example, particles generate fields that influence the movement of the particles and thereby change themselves.

In formal mathematical terms, the condition for autonomy in the case of discrete changes is as follows:

$$X_{n+1} = OP(X_n, P).$$

X_n represents the variables $x_1, x_2, \ldots . x_k$ of a network of operations at the time t_n; X_{n+1} is the state (value of the variables) that arises from the operation OP on the state X_n. P represents parameters $p_1, p_2, \ldots . p_k$ that are amplitudes that determine the weight of the individual operations. Because of autonomy, time is not an explicit variable. Without wanting to anticipate the following presentation, it has to be said that, in the case of social systems, the formal structure of the problem is identical but the symbols have different meanings: Variables are no longer quantitative and operations are no longer mathematical functions.

As each state is a direct outcome of the previous state, autonomous operation networks function recursively. The output of one operation becomes the input for the next one. Recursiveness is a direct consequence of autonomy: In an autonomous process, there can be no external specification of states-apart from the initial value.

Autonomy or recursiveness signify an operative detachment of an network from its environment. This is also called in literature *operational closure*.[8] By definition, operational closure occurs in a phenomenal domain if this domain is identical with the area in which the operations are defined. Insofar, the whole universe is operationally closed. But the more interesting question is, how can operational closure occur *within* an existing network of operations?

In a formal description, it is the observer who separates an area from the global operational context and views it as autonomous. However, the observer cannot proceed arbitrarily: The laws of nature must be satisfied; but they represent only very general principles. Hence, in a specific context, it is the observer's *epistemological interest* that determines which operations are selected and linked together from the set of possible operations - those that comply with the laws of nature - and which are ignored. This selection not only determines what is of "internal" significance, that is, whether, in the specific case, friction, thermal expansion, conductivity, and so forth can be ignored, but also

[7] Compare Krohn and Küppers (1992).

[8] See Varela (1981).

classifies "external" causes as being unimportant for the dynamics. Hence, operational closure is a selection process in which the epistemological interest of the observer functions as selection criterion. Generally, the order of the environment (boundary conditions) can serve as selection criteria for operational closure. In the laser, for example, it is the two mirrors that link the microscopic emission processes to the (macroscopic) light field.

Operational closure as a circular organization of different operations is a prerequisite for self-organization - causes and effects become mutually dependent and self-regulation can occur. But operational closure is not sufficient for the stable reproduction of a specific state of order. Another condition must be fullfilled: The different operations must have a specific relationship to each other. If (individual) operations are organized in this way, they will reproduce this organization through the (global) order that they create. In formal mathematical terms, the existence of stable, stationary solutions is required. Such solutions exist only for specific amplitudes of the various individual operations. For example, a threshold temperature gradient (convection) or a threshold pumping rate (laser) has to be overcome before the formation and decay of order compensate one another and create stable conditions. An internal variation of the amplitudes must ensure a "balanced" dynamic equilibrium of the system operations.

We have found two mechanisms that are a prerequisite of self-organization. One is a *selection* that transforms an open network of different operations into a closed one; the other is a *variation* that changes the amplitudes in the closed network in such a way that it reproduces itself. Under changing environmental conditions, the network can reorganize and produce a new order because of its nonlinear interactions across instabilities (deviation amplification). The strengths of the couplings in the network guarantee that this instability does not destroy the network but, at some point, comes to a standstill.

2. The Self-Organization of Social Phenomena

The fundamental problem in a theory on the self-organization of social communities (systems) is to show that individual actions lead to a rule system - a commonly held set of values, norms, and beliefs - that links individual actions to collective patterns of action that, in turn, stabilize this set of rules. Sociology offers two competing theories to describe the relationship between the individual and the community: Rational choice theory assumes that individuals possess action options and make choices that maximize own utility. Such individuals only contribute to the welfare of the community insofar as they can personally draw benefit from it. Collective goods such as general affluence, a clean environment, but also customs and traditions arise either as a by-product of activity that is otherwise directed toward individual utility or through constraint. If it is assumed that a social order exists, its continued existence is guaranteed by a complex interplay between sanctions and rewards.[9] At best, in social communities that are small enough, direct personal contacts generate a basis of trust, and specific mechanisms of social control are not necessary. In small communities, it is therefore rational for the individual to "cooperate voluntarily in the production of the public good of the social order."[10] This argument overlooks the fact that a basis of trust is already a collective good whose (voluntary) production has to be explained.

---

[9] Compare Olson (1965).

[10] Compare Taylor (1982, p. 94).

An almost complementary theory is provided by functional theories in sociology in which specific forms of collective activity are accepted as binding norms of individual behavior because of their usefulness for the community. An individual who acts rationally is "tied into a complex set of relations in which he must act trustfully because he has no choice."[11] Critics of such approaches claim that they provide no space for "subjective experience of individuals willing and choosing."[12] Individuals can act, not act, and perform a host of unexpected things. Other critics point to the arbitrariness with which social behavior can be assigned a function within a community. Only an additional selection theory can exclude this arbitrariness.[13]

Ludwig Fleck has indicated a way out of this dilemma: He has expanded the theory with a cognitive element (thinking style) that stabilizes and legitimizes the social community once it has arisen. Instead of proceeding from general beliefs and intentions, which subject social actions to the necessary restrictions and stabilize a community, the community is used to explain the generation of such rules. Fleck has developed the idea of the thought collective: A thinking style develops that defines what is held to be a reasonable question and a correct answer.[14] "Because of the general structure of the thought collective, the intracollective interaction of thoughts leads ipso sociologo facto - without reference to content and logical justification - to a reinforcement of the thought structure."[15] In contrast to classical functional theories, a community of individuals is placed at the beginning here. It develops the rules that structure its activity. This corresponds to the basic assumptions of a theory of self-organizing systems: A microscopic form of interaction of individual elements produces a macroscopic state of order, that then stabilizes its origins through feedback loops (that is, by a specific organization of the interaction). Hence, the question is: Which requirements have to be meet for a successful application of concepts of self-organization to social systems.

First of all, it is assumed that the basic operations of social systems are *social activity*. Typical forms of social activity are communication and cooperation. It is also assumed that this social activity does not occur at random but is guided by rules that can be labeled customs and traditions on the most general level. The analogy to the theories of self-organization in the natural sciences is then provided both by substituting social activities for operations and also by replacing natural laws with customs and traditions (in general terms, rules). In the natural sciences, the variables are relevant features (structures) of matter such as density, speed, pressure, temperature, and so forth. In social systems, these variables are features that characterize a concrete social system. For the science system, for which a detailed model of self-organization will be presented in the following, these would be, for example, experiments, theories, data, institutions, and so forth. The following table illustrates this analogy.

---

[11] Compare Douglas (1986, p. 31).

[12] Compare Douglas (1986, p. 32).

[13] Functional explanations in biology exclude arbitrariness by applying the concept of natural selection.

[14] Compare Fleck (1980).

[15] Compare Fleck (1980, p. 140).

Natural System		
Variables: Features of material systems such as:	*Operations:* Changes due to physical forces	*Elements:*
Pressure Temperature Speed Density	Attraction Repulsion Friction	Mass particles Electrons ....
Social System (Example: Science)		
Variables: Features of social sy-stems such as:	*Operations:* Social activities based on rules	*Elements:*
Experiments Data Institutions Theories Hypotheses	Operationalizing Measuring Interpreting Generalizing Justifying	Individuals

Table 1: Analogy Between Natural and Social Systems

3. Operational Closure

Social activities are governed by rules. This was the initial assumption here. The formation of a social system with a specific pattern of social activities at first requires operational closure. Here as well, external boundary conditions form the selection criteria that are responsible for this. The first boundary conditions that come to mind are the various institutionalized rule systems[16] in society: laws, statutes, contracts, and so forth. More generally, it is the expectations of the social environment that select specific forms of social activity as boundary conditions. Without these, social activity, which can be triggered at random, would simply subside again.[17]

In functional systems, it is the environments' expectation of a specific product that serves as a boundary condition to "control" the self-organization within the subsystem. Such typical products are scientific knowledge in the science system, "just" rulings in the legal system, normative decisions in the political system, and so forth. If these external expecta-

[16] Implicit rules, which are typical for social systems of short duration and with less formal forms of cooperation, give way at some point of time during the history of the system to explicit forms of institutionalized rules that fix and stabilize this social system more permanently. This is because institutionalized rules free individuals from illegitimate claims and false expectations.

[17] Even on the lowest level, the level of social interaction between two individuals, expectation plays a decisive role as a selection criterion. A social interaction lasts for a certain length of time only when ego does not disappoint the expectation of alter ego (and the same applies in reverse).

tions on performance function as external boundary conditions and have such an impact on social activity that specific continuations are selected in preference to others (production of an asymmetry), then certain *rules* of operation arise conditions that ensure that one pattern of action separates itself from others and then either "functions" or collapses. Such rules determine the organization of operations within social system.

4. Research Teams as an Example of Social Self-Organization

The rest of this paper is concerned with the application of self-organization to a specific social system, namely science. Within this system, research teams are all those social systems that are the smallest social units responsible for the production of new knowledge in the science system.[18] Research is defined as activities that are directed systematicaly toward the production of new knowledge.

As a rule, research teams are embedded in more extensive institutional systems such as laboratories, institutes, or, completely generally, formal networks of cooperation. Although these function as *boundary conditions* to restrict the interactions in research teams, they finally do not determine the interactions. However, they permit the selection of task--specific interactions and their combination or integration.

Nonetheless, these external boundary conditions grant research teams a degree of freedom in finally determining the way in which their research targets can be achieved. By exploiting this freedom - in the complex network of task-specific interactions, certain forms of interaction are considered to be more important than others - specific interaction rules are formed that are considered to be necessary to achieve the selected target. It is decisive here that this set of rules is reproduced by the continuation of interaction and thus sets up the research team as a permanent social system.[19]

Research teams produce new knowledge. That is their system-specific function within the science system. This production of knowledge is a process of self-organization that will be inspected more closely in the following. The hypothesis is that new knowledge is a stable, stationary state (eigensolution) of a recursive and operationally closed research process. This means that new knowledge is always gained when the research process reproduces itself in the sense that new points of view - in the language of self-organization: "perturbations" - do not lead to the reformulation of statements (findings) that have already been made.

Describing system-specific performances as eigensolutions of a process of self-organization calls for the introduction of appropriate *variables* and the definition of system-specific *operations* as mechanisms of their change. Here, variables are understood as changeable quantities, that is, objects, phenomena, or also facts that are typical for the institutions under consideration and the processes that unfold in them.

On the basis of empirical studies and the philosophy of science, it can be stated that research findings in the science system are determined by the coupling of six variables that can each take on different "values." This does not mean quantitative values but different qualities of each variable.

---

[18] For a detailed discussion, see Krohn and Küppers (1989). For methodological reasons, research performed by individuals is not considered in this study.

[19] Alongside environmental expectations, this set of rules also encodes the expectations of the individual members of the research team. Elsewhere, we have labeled this the cognitive-emotional group matrix (Krohn & Küppers, 1989, pp. 37ff.).

VARIABLES	VALUE DOMAIN
PROCEDURES	Experimental designs, simualtion models, numerical codes, field studies, etc.
DOCUMENTS	Data printouts, protocols, film recordings, interviews, etc.
DATA	Numerical values, graphs, empirical examples, etc.
PATTERNS OF DATA	Correlations, redularities, causal relationships, interpolations, etc.
THEOREMS	Derivations, confirmations, rejections, etc.
HYPOTHESES	Statements, questions, etc.

Table 2: Variables in the Research Process

Each scientific finding contains a selection of suitable "values" of the individual variables in which this suitability results from their mutual dependence: The values have to form an eigensolution of the dynamic process of "knowledge production." As one single research team cannot perform a complete estimation of this suitability in advance, first of all, initial values are selected that are considered to be plausible although they are finally arbitrary. Through recursive operations, it is then ascertained whether these values continue to function as a stationary solution, require slight modifications, or need to be replaced completely.

In addition, the functional dependencies of the variables have to be defined through which the knowledge production is described and presented as a closed circular process. Typical for the structure of a finding, particularly, a contribution to empirical research, is the coupling of two relatively independent subprocesses: the integration of conceptual (theoretical) and data-generating research activities. This coupling is performed by two procedures that can be labeled, in general terms, as the *operationalization* of rules (laws, theories, etc.) and the *interpretation* of actualities (facts, data). In each production of knowledge, the rules (scientific laws) have to be operationalized in a more or less explicit

DERIVATION (of theorems)	=	Hypotheses
INTEGRATION of (patterns of data)	=	Theorems
INTERPRETATION of (data)	=	Patterns of data
SELECTION/EVALUATION of (documents)	=	Data
APPLICATION of (procedures)	=	Documents
OPERATIONALIZATION of (hypotheses)	=	Procedures
DERIVATION (of theorems)	=	Hypotheses

Table 3: Operations of the Research Process

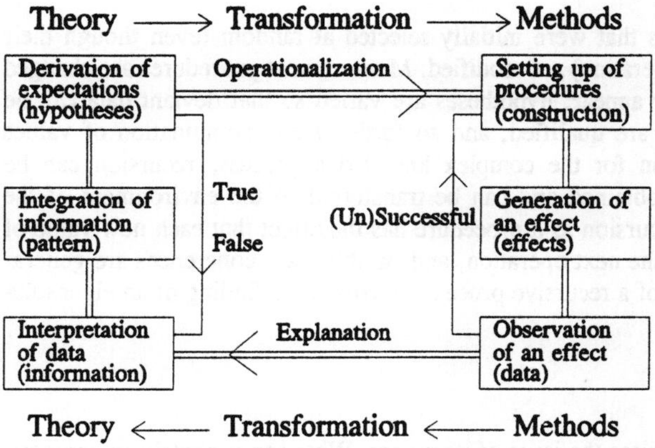

Fig. 1 The structure of the knowledge operation. The left-hand suboperation (argumentation) utilizes the dichotomy of true/false; the right-hand (effective action): successful/unsuccessful.

way so that they fit a new case, a new problem. Vice versa, the facts of the problem (data) have to be interpreted until they correspond to the rules. Hence, operationalization and interpretation are paths of modification. These modifications have to be performed until - initially in the eyes of the observers, but in expectation of agreement from the environment - an informational structure emerges that is viewed as a satisfactory solution to the problem. The construction of such a solution can be labeled the *construction of an eigensolution*. An eigensolution accordingly consists of a stable coupling of interpretations of available data, hypotheses derived from this, appropriate procedures for testing these hypotheses, and the data thus generated. Because of its circularity, this process can also be entered at another point.

This can be summarized as follows:

Hypotheses = Derivation (Integration (Interpretation (Selection/Evaluation (Application (Operationalization of (*hypotheses*))

or:

$$H = OP(H)$$

in which H represents hypotheses and OP the network of operations in the process of knowledge production. H is the eigensolution that is being sought for the complex operation of knowledge production. Depending on the specific epistemological interest, data, procedures, and so forth can also be constructed as eigensolutions.

Hence, the construction of an eigensolution consists in finding a stationary state for the system of coupled operations determining the complex process of knowledge production. This requires, on the one hand, decisions on the relative weighting of the individual operations ("internal variation," i.e., theoretical arguments can be seen as more important than methodological ones); while, on the other hand, values for the various variables must be selected in such a way that they do not change any further when the procedure is repeated.

In a recursive procedure, values that were initially selected at random (even though their plausibility was taken into consideration) are modified: Measurement procedures are changed so that unwanted data no longer appear; hypotheses are varied so that deviant data can be integrated; relations to theories are qualified; and so forth. If one combination of values finally provides an eigensolution for the complex knowledge process, recursion can be aborted: A research finding is obtained and can be transferred to the environment of the research group as a product. Recursion as a procedure has the effect that each new value of a variable becomes the input of the next operation, and, in this way, coherences are generated. In addition, the stationarity of a recursive process signalizes the finding of an eigensolution.

5. Conclusions

The presentation of this model raises the issue of its power. What can it explain in comparison to conventional theories of science studies? In general terms, this approach requires only a few a priori assumptions because it has a greater power of self-structuring. Self-organizing systems generate structure, and problematic decisions on what belongs or does not belong to science do not have to be made as an apriority of theory formulation but are resolved by the dynamics of the system. In specific terms, I can see new constructive possiblities in three places:

1. The problem between system and environment presents itself in a new form. While conventional approaches are based on the assumption of factual alternatives between scientific autonomy and heteronomy, between internal and external determinants of development, this model does not consider this dichotomy to represent two mutually exclusive alternatives: Science is neither autonomous (self-regulated) nor heteronomous (externally controlled). Instead, it can either increase or weaken its heteronomy through its autonomy. Being receptive to social needs definitely does not mean external control but is the best protection against illegitimate demands.

2. Self-organizing systems are rule forming: Organized structures in the macroworld emerge from unorganized microdynamics. The rules of this organization are not set in advance but arise through appropriate feedback from the effects to their causes in the system. The central theoretical category is the formation of eigensolutions in operational closed systems. Hence, methodologically, it does not permit rules to be set in advance, as in the functional theories of science studies or in the theory of critical rationalism - for example, rules and standards of scientific rationality. Science is at its most productive precisely where it violates its own rules.

3. The theory of self-organization permits the modeling of the social construction of scientific knowledge because the mechanisms of construction can be specified. While so-called ethnomethodological constructivism restricts itself to inferring the presence of social mechanisms from the absence of specific rational mechanisms for the construction of a scientific fact, without admittedly being able to specify these social mechanisms, the framework of self-organization can be used to demonstrate how scientific rationality is itself a social construction. While the research process proceeds in an anarchic manner, scientific rationality comes into being as a product of the feedback of those activities that are directed predominantly toward the spread, application, communication, and justification of research findings.

References

Douglas, Mary, 1986: How Institutions Think, Syracuse: Syracuse University Press.

Fleck, Ludwik, 1980: Entstehung und Entwicklung einer wissenschaftlichen Tatsache. Einführung in die Lehre vom Denkstil und Denkkollektiv, Reprint Frankfurt am Main: Suhrkamp.

Krohn, Wolfgang, Günter Küppers, 1989: Die Selbstorganisation der Wissenschaft, Frankfurt am Main: Suhrkamp.

Krohn, Wolfgang, Günter Küppers, 1992: Die natürlichen Ursachen der Zwecke. Kants Ansätze zu einer Theorie der Selbstorganisation, in: Wolfgang Krohn, Hans-Jürgen Krug, Günter Küppers (eds.): Konzepte von Chaos und Selbstorganisation in der Geschichte der Wissenschaften. Jahrbuch Selbstorganisation Bd. 3, Berlin: Duncker & Humblot. Pp. 31-50.

Krohn, Wolfgang, Günter Küppers, Rainer Paslack, 1987: Selbstorganisation - Zur Genese und Entwicklung einer wissenschaftlichen Revolution, in: Siegfried J. Schmidt (ed.): Der Diskurs des Radikalen Konstruktivismus, Frankfurt am Main: Suhrkamp. Pp. 441-465.

Luhmann, Niklas, 1984: Soziale Systeme. Grundriß einer allgemeinen Theorie, Frankfurt am Main: Suhrkamp.

Olson, Mancur, 1965: The Logic of Collective Action: Public Goods and the Theory of Groups, Cambridge, Mass.: Harvard University Press.

Taylor, Michael, 1982: Community, Anarchy and Liberty, Cambridge: Cambridge University Press.

Varela, Francisco J., 1981: Autonomy and Autopoiesis, in: Gerhard Roth, Helmut Schwegeler (eds.): Self-Organizing Systems. An Interdisciplinary Approach, Frankfurt am Main / New York: Campus. Pp. 14-23.

Weise, Peter, 1990: Der synergetische Ansatz zur Analyse der gesellschaftlichen Selbstorganisation, in: Ökonomie und Gesellschaft. Jahrbuch 8: Individuelles Verhalten und kollektive Phänomene, Frankfurt am Main / New York. Pp. 12-64.

The Significance of Nonlinear Phenomena for the Investigation of Cognitive Systems

P. Kruse and M. Stadler

University of Bremen, Institute of Psychology and Cognition Research, D-2800 Bremen 33, Fed. Rep. of Germany

Abstract: There are good reasons for understanding the brain as a complex self-organizing system. Despite the fact that nonlinear phenomena are an indicator for self-organization they are only rarely reported in psychological research. Some possible explanations for that are discussed: The preference of homeostatic models, the lack of dynamic measurement and the tendency to reduce complex cognitive phenomena to elementary processes. The synergetic approach presents a model of brain-mind interaction. Synergetics postulate pattern recognition as an analogue to pattern formation. In that view cognitive processes are directly represented by the macrodynamics of the brain. In the article first nonlinearities in complex learning tasks (learning plateaus) are presented and interpreted as the emergence of a hierarchy of self-organizing order parameters. The second example of important nonlinearities in cognitive systems are multistable perceptions. Here we find that minimal semantic influences may affect the perceptual dynamics during the instable phase. A further field of nonlinearities in perception is represented by the hidden potential landscapes of homogeneous areas. Using the method of serial reproduction an attractor structure is revealed which is again sensitive to semantic influences. Finally it is demonstrated that even complex human actions may be analyzed as an attractor driven self-organizing process explaining sudden behavioral changes.

1 Nonlinearity in psychological research

The existence of nonlinear phenomena in the behavior of a system is a necessary condition for and a strong hint at self-organization processes. All the fascinating examples of self-organization discovered and elaborated in the fields of physics, chemistry, and biology during the last three decades started with irritating and unexpected observations of spontaneous and sudden changes in the behavior of natural systems [1]. Inspite of the basic theoretical assumption of continuity in nature (natura saltum non facit) the complex dynamics of self-organizing systems is able to jump from one to another stable state of order. These non-equilibrium phase transitions are caused by autocatalytic amplification of elementary fluctuations which enables the emergence of new macroscopic states of order in self-organizing systems. From a microscopic point of view the systems are unpredictable. In phases of instabiliy they are open to minimal influences causing maximal behavioral effects.

The human brain is a system which seems to be able to produce a nearly endless variety of ordered macroscopic states and which is probably the most complex system in nature. There is much physiological evidence that the brain has to be understood as a self-organizing system [2]

Springer Series in Synergetics, Vol. 62 **Interdisciplinary Approaches to Nonlinear Complex Systems** – Eds.: H. Haken and A. Mikhailov

and theoretically this is an obvious and by no means new idea. Already in 1925 Wolfgang Köhler the important protagonist of gestalt theory wrote: "The somatic processes underlying static visual fields are stationary equilibrium distributions developed from the inner dynamics of the optical system itself" [3]. Though the holistic approach of gestalt theory was constricted by the concepts of linear thermodynamics of those days, gestalt theory can be understood as a precursor of the modern theory of self-organizing systems [4]. In contrast to this early conceptualization of a theory of self-organization in brain and cognition, nonlinear phenomena are only rarely reported in the psychological literature. Which are the possible reasons for the discrepancy between the theoretical evidence of self-organization processes in cognition and the lack of empirical reports of nonlinear phenomena in psychological research?

One basic reason may be that conceptualizations in psychology are directed up to now more or less strictly towards homeostatic modelling. This preference reflects the fact that the least questionable goal of a living system is to survive. Life of a biological system is strictly bound to homeostasis. In homeostatic modelling the preservation of a clearly defined stable state is the basic mode of operation of the system. Unpredictable spontaneous changes can only be regarded as a breakdown of normal functioning. Therefore also in cognitive organization nonlinearities and phase transitions seem to be marginal phenomena and are not in the focus of interest. Yet, a purposeful reanalysis of already existing empirical results of psychological research might bring up more nonlinear phenomena as found at first sight.

A second closely related reason can be seen in the relative narrow bands in which phase transitions occur. In the behavioral space of a biological system phase transitions are necessarily limited to well-defined border conditions. Regimes of linear system behavior are the rule and nonlinearities have to be the exception to guarantee a stable basis of living and action. For cognitive functions like perception or thinking and from a phenomenal point of view the problem is even further intensified. To guarantee a stable basis of action a phase transition from one stable state to another - if existent at all - has to be very quick and no or not much conscious capacity should be vasted on recognizing the process of cognitive order formation. Therefore by principle it will not be easy to detect and measure phase transitions in psychological experiments. Psychological measurement once more is critically dependent on highly indirect methods.

Like the second also the third reason contains a theoretical and a methodological aspect. In experimental psychology there exists a tendency to try to reduce the complexity of natural cognitive phenomena by analyzing elementary processes under very restricted conditions [5]. On one hand this is due to the requirements of the experimental methods used and on the other hand it results from the assumption that the complexity may be rebuilt by connecting the elementary findings. In some cases the research is even limited by the technical equipment available. This was f.i. the case in research on visual perception. For a long time in nearly all experiments only static displays were used before event perception became the basic level of analysis [6]. To observe the full dynamics of a complex living system it is necessary to look for experimental methods which allow a systematic analysis of behavior without reducing and constraining the system too much. The appearance of self-organization characteristics needs the possibility of an unconstrained development of the system behavior. The system must be able to follow its own inner dynamics more or less freely.

2 Synergetics of cognition

The revaluation of the significance of nonlinearities in empirical psychology is the starting point and the consequence of a self-organization theory of cognition. In the theoretical framework of synergetics [7] the first step of analysis is to demonstrate the existence of phase transitions in a complex system. If such phase transitions can be shown a number of concrete theoretical expectations have to be satisfied to categorize the phenomenon as consequence of a self-organization process. For the spontaneous reorganization a certain control parameter has to be defined which releases the sudden transition from one to another stable state of order when continuously enhanced. Approaching the point of change by gradually increasing or decreasing the control parameter the system behavior should show a tendency to persist in the previous stable state (hysteresis). Before the phase transition an autocatalytical destabilization of the system appears. This destabilization is manifested by critical fluctuations and by a critical slowing down of the innersystemic tendency to conserve the existing stable state of order. In synergetics the autonomous reorganizations are explained by the appearance of different modes of behavior competing till one of them predominates the others by slaving the behavior of elementary components of the system. This predominating mode is called an order parameter. The self-organisation process is characterized by a certain circular causality between the microscopic and the macroscopic level of system behavior. The macroscopic stable states of order emerge from and organize the microscopic interactions of elementary components of the system.

This idea of a circular causal interaction between micro and macro processes in self-organizing systems is a powerful metaphor for modelling the mind-brain-relationship [8]. The micro processes of the nervous network give rise to a macroscopic collective process, for instance to a temporal synchronicity distributed over more or less distant areas of the brain. This collective process represents the order parameter which organizes the microscopic activities in the neural network (figure 1). One of the main results of the synergetic approach is the obvious analogy between pattern formation and pattern recognition [9]. In the highly developed biological system of the brain pattern formation can be understood as being identical with pattern recognition. This assumption represents an important step towards an empirical handling of the mind-brain-problem. If the macroscopic order parameter, which has emerged from the

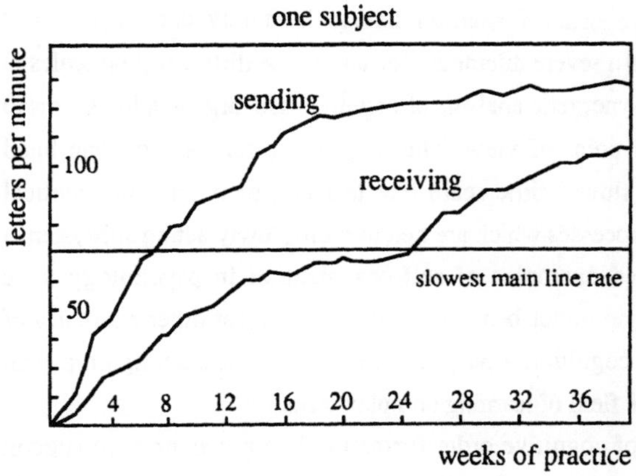

Fig. 1: Micro-macro-relation in brain-mind-dynamics

elementary activity, represents and governs this activity, it may at the same time represent the cognitive process which influences the behavior of the organism.

Following the assumption that pattern formation is identical to pattern recognition, some basic problems of the mind-brain-relation can be overcome. The dynamics of perception, thinking and memory need not be reduced to elementary brain processes. The cognitive dynamics may be represented directly by the macrodynamics of the brain. Cognitive phenomena can be used as a methodological window for observing and understanding brain activity. In this view cognition is not an epi-phenomenon of brain activity but its result and its organizing factor. The emergent psychological quality of meaning is necessary for the organisation of the complex neural activities taking place in the human brain.

3 Some methodological problems

Assuming that cognitive order formation is a process of self-organization some fundamental methodological problems arise which have to be considered in psychological research. Any self-organization process is intimately connected with instability before and during the phase transition from one stable state of order to another. In contrast to the expected instability in the process of cognitive order formation, it is one major characteristics of the world of experience to be stable. At first sight instability is only experienced in ontogenetic processes like learning and not i.e. in the actualgenetic process of perception. The neural processes of self-organization underlying the cognitive order formation act on a time scale far beyond conscious realization. As already mentioned, from an evolutionary point of view it does not make any sence to vaste cognitive capacity on recognizing the process of order formation. To guarantee a stable and permanent basis of action cognitive order formation has to be fastly converging. For experimental methodology this leads to a severe dilemma. Because of the different time scales in most cases a direct proof and a direct synergetic analysis of cognitive self-organization seems to be impossible from a psychological point of view. The only exceptions are on one hand cognitive processes which act on a slower time scale like learning or other ontogenetical developments and on the other hand processes which are tied to more slowly acting subsystems. Consequently the first fully analyzed examples of self-organization in psychology were published for the organization of human motor behavior [10]. In search for other examples of non-equilibrium phase transitions in cognition a purposeful reanalysis of existing empirical results should be most promising in the field of learning complex tasks.

For fast converging processes of cognitive order formation like perception we suggest three main experimental strategies to solve the described dilemma. To perform a synergetic analysis in the way outlined above one can (i) refer to a faster time scale by direct registration of the neural dynamics underlying cognitive order formation i.e. by performing correlative EEG studies, (ii) analyze the process of cognitive order formation in situations where the cognitive system is in an instable phase i.e. by using the phenomenon of cognitive multistability, (iii) slow down the process of cognitive order formation by externalizing intermediate results i.e. by using the experimental technique of serial reproduction (Bartlett scenario, 11).

In the next chapter some ancient experimental results of research on learning will be reanalyzed from the synergetic perspective. Thereafter some experiments on perception will be presented which were designed on the basis of the last two strategies mentioned. Finally an example for a sudden change between behavioral patterns in a complex social situation will be given which can be described in terms of a non-equilibrium phase transition.

4 Autonomous order formation and learning

Theoretically learning can be understood as a process of self-organized order formation. The constant learning effort of the learner while difficulty of the learning material is increased can function as a control parameter and his achievement is the result of emerging cognitive structures. If learning is a process of self-organization it has to be expected that the achievement curve of the learner does not lineary increase over time. The learning curves should show phases of linear increasing, phases of stagnation, and phases of significant sudden improvement of performance. During the linear increase a cognitive structure is optimized and transferred into performance. After a maximal optimization a phase of stagnation follows where a destabilization of the existing structure starts. The phase of sudden improvement is initiated by a new emerging cognitive state of order. Consequently as a first step three empirical expectations can be derived. (i) When measured over a sufficiently long time and with a nearly constant learning effort and a linear increase of task difficulty, learning processes will show characteristic nonlinearities in the

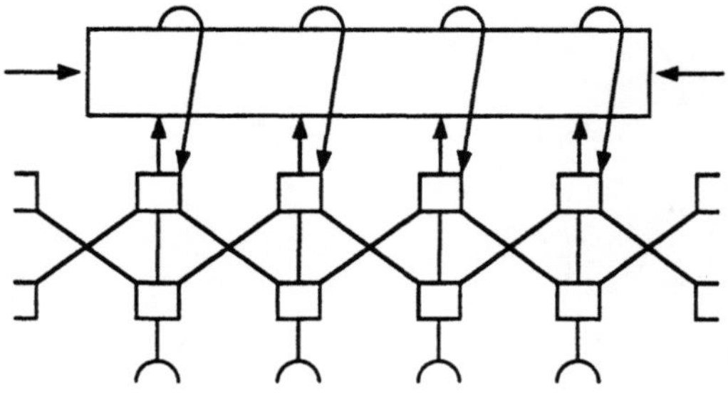

Fig. 2: Nonlinearities in the learning curve

performance of the learner (phase transitions). (ii) The stability of some parameters of the performance will breakdown at the end of each phase of stagnation (critical fluctuations). In contradiction to naive expectation one conrete hypothesis is that at the end of the highly optimized application of one strategy the rate of errors increases. (iii) Again in contradiction to naive expectation the learner will show also a significant increase in sensitivity to disturbances at the end of the phase of optimalized application (critical slowing down).

The existing of nonlinear, stepwise development of performance in learning has been reported very often as a part of learning experience. But there are almost no published empirical results for such learning plateaus. Mostly learning processes were investigated for short time intervals with material being not very complicated . One interesting exception is a study of Bryan and Harter published in 1897/99 [12]. The authors investigated a very complex task - the aquisition of the telegraphic language - by different subjects over up to 40 weeks. Task difficulty and performance could be easily quantified by the number of letters the subjects were able to send or to receive. As expected Brian and Harter found typical nonlinearities in performance over time of learning but intrestingly only for the receiving of telegraphic language (figure 2). In the situation of sending the meaning of the words to be translated is always given and the number of submitted letters is only restricted by assoziation of the different alphabets and by the degree of automatisation of movement. Therefore the learning curve for sending reflects mainly a single process of optimization with a phase of nearly linear increase reaching asymptotically an optimum of performance. This is the traditional form of the one-dimensional learning curve. But concerning receiving the situation is totally different. During receiving meaning always has to be generated from the incoming impulses. Performance is critically dependent on the cognitive strategy used and the cognitive order built up. Receiving is far more complex than sending. The subjects of Bryan and Harper reported that the learning plateaus and the following sudden increases of performance in receiving are due to the change from letter and word understanding to the direct understanding of whole sentences. Only after years of further practice some experts pass another learning plateau. They describe the next sudden step in performance as the ascent

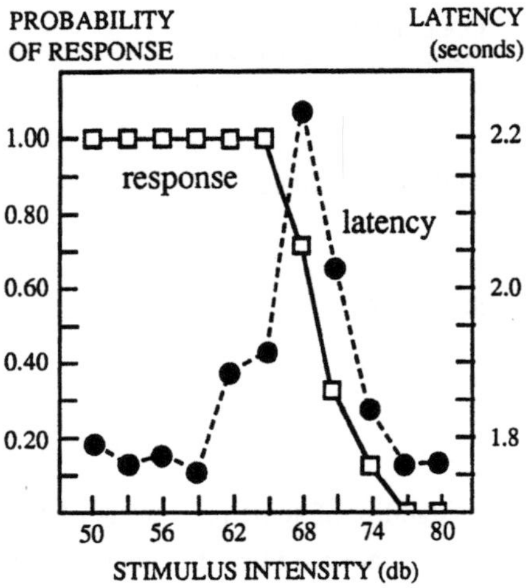

Fig. 3: Stimulus generalisation and response latency

from drudgery to freedom. This step may be comparable to the step from translating a language to thinking the language. In the view of synergetics the different steps result from higher states of cognitive order emerging in the learning process. The different forms of meaning (from letter and word understanding to the understanding of sentences) and of language habits (from translating a language to thinking the language) act as order parameters. They emerge from and govern the process. For accepting the hypothesis of self-organization underlying such learning processes it is necessary to show the existence of critical fluctuations and critical slowing down. This has to be done in future research.

But before switching to own empirical work on perception some other existing results shall be discussed out of the synergetic perspective. These results are part of a study of auditory generalization following multiple-response discrimination training published by Cross and Lane in 1962 [13]. In one of their experiments subjects were trained to distinguish between two clearly seperated intensities of a sound (56 db and 74 db) by humming with two different levels of pitch as a response. When the subjects had learned to react correctly the sounds were given with a whole range of intensities (from 50 db up to 80 db in 3 db steps) in a random order to measure the stimulus generalization. As shown in figure 3 the responses of the subjects show a very broad generalisation with a sharp decline of probability for both response behaviors and a dramatical increase in response latency during this decline. Looking at the results we may suggest that there is an hysteresis effect when gradually approaching the intensity range of the decline from both sides of the scale. The data seem to indicate a phase transition between two attractor states established during the discrimination learning. The dramatical increase in

Fig. 4: Instable visual pattern

response latency in the range of the sudden change between the two response categories can be interpreted as a critical slowing down of the process.

5 Elementary dynamics and perceptual instability

In view of synergetics inspite of the apparent stability of our daily perceptions, instability and multistability are basic features of all perceptual processes. The perceptual exception of multistability is a methodological window to the basic process of autonomous order formation which usually is too fastly converging to be consciously recognized [14]. During the oszillation between different stable states the dynamics underlying perceptual order formation can be investigated. In perception research there are many of examples of such spontaneous and sudden reorganizations reported and analyzed. Figure 4 shows for instance a pattern where the fluctuating activity in the visual system on search for stability can be observed directly. There are obviously different modes (collective processes) which compete for some time and which give the whole pattern a dynamic appearance. None of the modes is able to predominate as an order parameter for a longer period. No pattern is stabilized for more than a few seconds.

Multistable patterns like the Necker-cube, Rubin's vase/ face-picture, the Maltese-cross, the rabbit/dog-pattern and many others usually lead to perceptions with two or more predominating stable states which alternate periodically. Figure 5 shows the underlying potential of such a reversible process, as it is used for model calculations by Ditzinger and Haken [15]. In own experiments we preferred dynamic displays to analyze multistable behavior because in motion perception the systematic variation of a control parameter is easier than in static pictures. Additionally ambiguous apparent movement (AM) patterns are very reliable in their dynamic behavior, the reversion process is well defined for perceivers, and they often allow more than two stable states [16]. As stimulus we used the stroboscopic alternative movement (SAM, see figure 6) and the circular apparent movement (CAM, see figure 7). For these patterns the oscillation between the movement alternatives shows the phenomenon of hysteresis as it is predicted in the synergetic model of phase transitions.

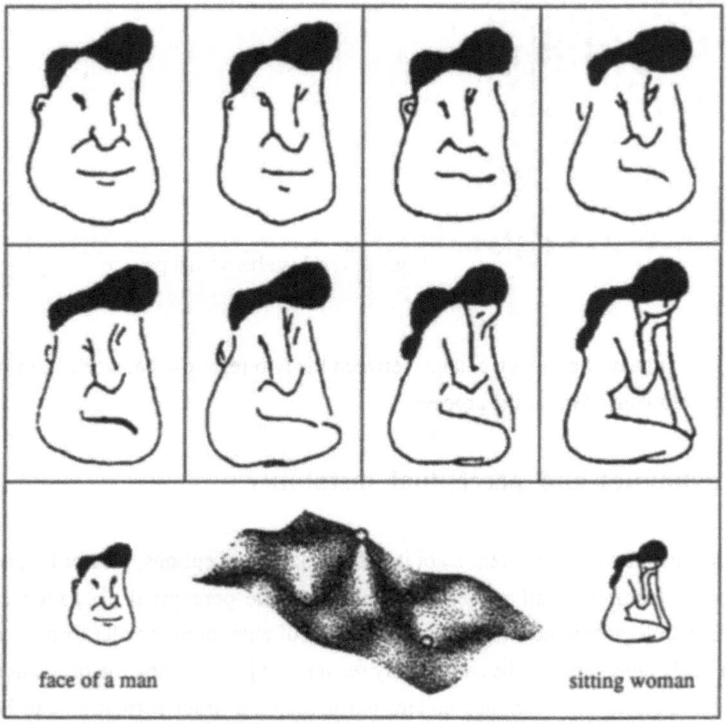

Fig. 5: Man`s face to girl reversion and potential landscape

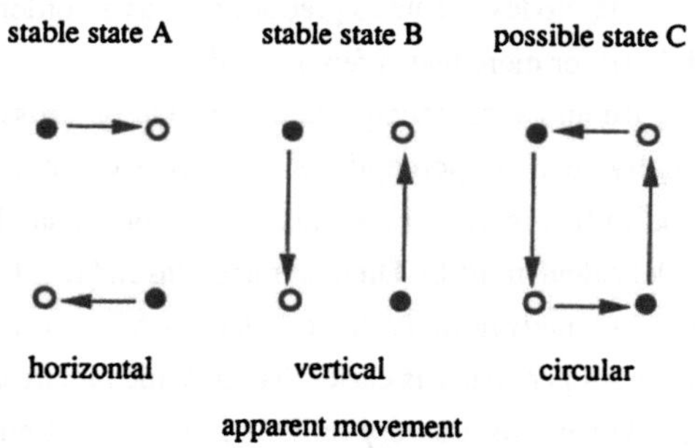

Fig. 6: Stroboscopic alternative movement (SAM)

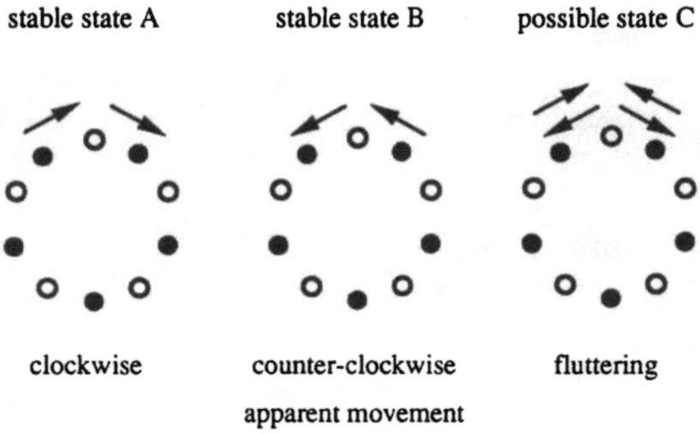

Fig. 7: Circular alternative movement (CAM)

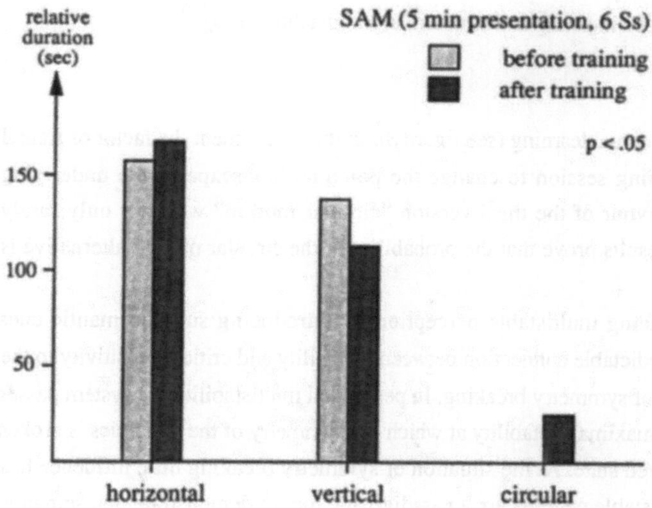

Fig. 8: Changing the probability of apparent movement alternatives in CAM via perceptual learning

In a variety of experiments we were able to demonstrate that it is possible to change the degree of stability or instability of such multistable patterns by contextual and semantic influences. The potential landscape underlying the dynamics of the reversion process can be altered by introducing gestalt factors like common motion or figural identity, by perceptual learning, or even by very subtle semantic cues, i.e. initially multistable patterns can be made monostable [17].

In one of our experiments we were able to show that the mostly bi-stable SAM can be further destabilized by enhancing the probability of a third theoretically possible perceptual

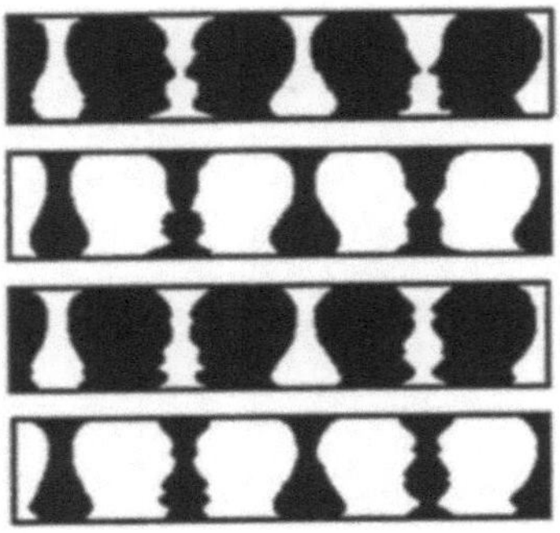

	vase	face
row 1	0%	100%
row 2	20%	80%
row 3	47%	53%
row 4	61%	39%

(n = 20 each group)

Fig. 9: Interaction between meaning and structure in multistability (19)

alternative to appear via perceptual learning (see figure 8). In this experiment the factor of figural identity was used in a training session to change the potential landscape of the underlying dynamics of the SAM in favour of the third version "circular motion" which is only rarely perceived spontaneously. Results prove that the probability of the circular motion alternative is enhanced by training.

The aspect of influencing multistable perception by introducing subtle semantic cues supports the theoretically predictable connection between instability and critical sensitivity to the initial conditions in the case of symmetry breaking. In perceptual multistability the system passes again and again the point of maximal instability at which the symmetry of the dynamics is broken in favour of one stable ordered state. At the situation of symmetry breaking little influence has great effect. Therefore multistable patterns are a paradigmatic tool to demonstrate that semantic cues are able to influence macrodynamic brain processes. There are many examples in perceptual research showing top down influences from meaning to structure and even to basic sensory qualities in the research program of the so-called "new look" of the late forties and fifties [18]. One and the same green-brownish colour, for instance, is judged by many subjects more brown, if it is exposed in the form of a horse and more green if it is exposed in form of a leaf. Figure 9 shows that there is an interaction between the meaning and structure in multistability. The structurally more attractive (gestalt factors of symmetry and proximity) vases only predominate in the figure ground pattern, when the meaning of human faces cannot be attributed.

For the investigation of semantic effects on the perceptual dynamics, we used again the stimulus patterns SAM and CAM. If the CAM, for instance, is composed of arrows pointing in an anti-clockwise direction, this direction is preferred in the bifurcation situation. At first sight nearly all subjects see an anti-clockwise rotation of the apparent movement although usually the

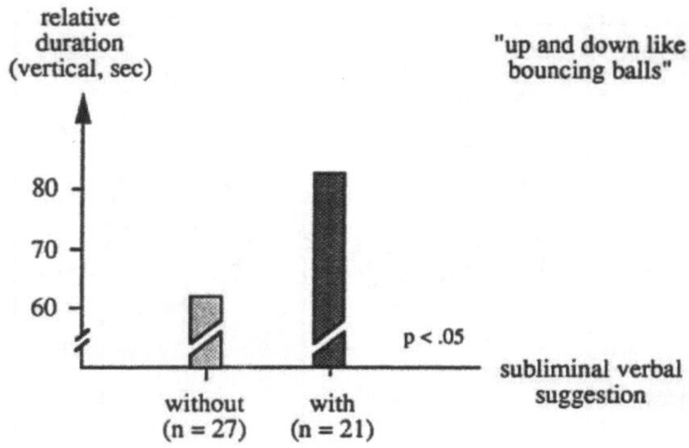

Fig.10: Changing the probability of apparent movement alternatives in SAM by subliminal suggestion

clockwise rotation is preferred. Even subliminal verbal suggestions, e.g. "up and down like bouncing balls", given to the subjects below the threshold of conscious recognition (masked by music) during presentation of the SAM has a significant effect on the relative duration of the movement alternatives (figure 10).

The fact that meaning is able to influence the structure of brain processes, is predicted by the synergetic model of mind-brain-interaction. Further research should investigate the properties of macrodynamic brain processes during the instable and stable phase. First experiments in that direction show that EEG recordings during the perception of the SAM are significantly different from recordings during the perception of nearly identical stable AM patterns [20].

6 Attracting field forces and serial reproduction

Another interesting method for the analysis of cognitive order formation has been invented by Frederic C. Bartlett [21], the method of serial reproduction. This method is an adaptation of the mathematical priciple of iteration which is of special interest in chaos theory and fractal geometry [22].

Iteration is a simple feed-back process in which the same operation is carried out repeatedly, the output of one calculation or step being the input for the next one. Depending on starting conditions such a process tends to or produces different states of order. These attractor states are inherent in the specific operation on which the iteration is based. Bartlett used the method of serial reproduction to show that there is systematic spontaneous change in simple drawings when reproduced in a series of iterations. He presented i.e. the picture of an owl to one person, the reproduction of that person to another person and so on (see figure 11). The resulting series

Fig.11: Serial reproduction of the drawing of an owl

strongly remind of the swiching from one perceptual alternative to the other in multistablity (figure 5). The method of serial reproduction produces instabilities and phase transitions between different attractor states. Like multistability serial reproduction can be used to analyse the dynamics underlying cognitive processes. Serial reproduction makes visible the attractor states inherent in the cognitive operation. In synergetics the dynamics of a self-organizing system is described by potential fields. The serial reproduction is a way to measure attracting field forces to evaluate such potential landscapes in the process of cognitive order formation. Thereby hidden nonlinearities can be demonstrated.

On the basis of serial reproduction we investigated the old phenomenon of the wandering dot [23]. A dot on a blank piece of paper seen for a few seconds changes it's position on another blank paper after reproduction. This reproduced dot is again reproduced by another subject and

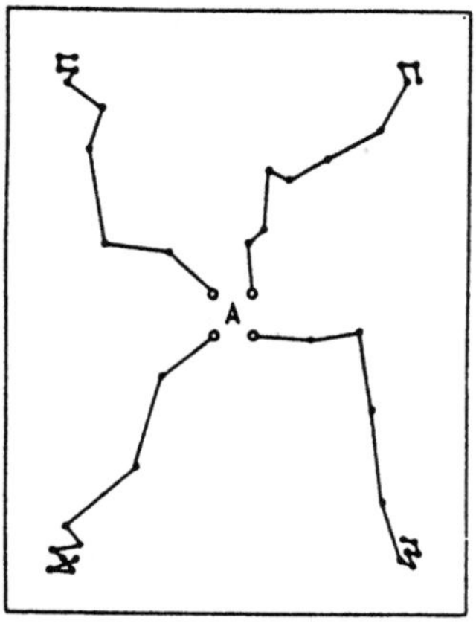

Fig.12: The phenomenon of the wandering dot

so on. Systematic analysis of the phenomenon shows typical trajectories of wandering dots starting from different positions A in the middle of the area. The dots first take small steps, then longer steps and then diminish their steps again until they have found a more or less stable position near one of the corners. The trajectories look as if there is an invisible potential gradient distributed over the area which causes the dots to make these particular movements (figure 12). Obviously the wandering dots show a nonlinear behavior on this virtual potential gradient and such a behavior seems to be in contradiction to the linear texture gradients distributed over surfaces as described by J.J. Gibson [24] and the ecological school.

Attempts were made in our laboratory to measure the underlying nonlinear potential fields [25]. For this purpose first the displacement vectors (single point presentation) were collected for a 21x29 dot-position-pattern from 10 Ss (figure 13).

In a next step the raw vectors of all individual experiments were subjected to a vectorial analysis procedure by which the sources of each vector are integrated over the whole pattern and the circular potential is divided from the gradient potential. Figure 14 shows the calculated vector field of the averaged data of 10 Ss. There is a very regular distribution of the vectors, showing the nonlinearities of the field, i.e. the bifurcation saddle in the middle and the attractors near the four corners.

In Figure 15 the gradient potential is presented as a landscape over which the dots move like spherical particles. Figure 16 is the result of a model calculation of these movements, which resembles very good the empirical data of figure 12.

By the experiment the hidden nonlinear structure of a homogeneously stimulated perceptual field is manifested. Stabilities and instabilities in this field are represented by bifurcation areas (repellers) and attractor areas. Comparably to multistability one can try to modify the shape of the potential landscape by introducing cognitive factors like meaning. Figure 17 shows the changes in potential landscape when the dots are replaced by arrows pointing

Fig.13: Matrix of stimulus positions for single dot reproductions and the actually measured displacement vectors

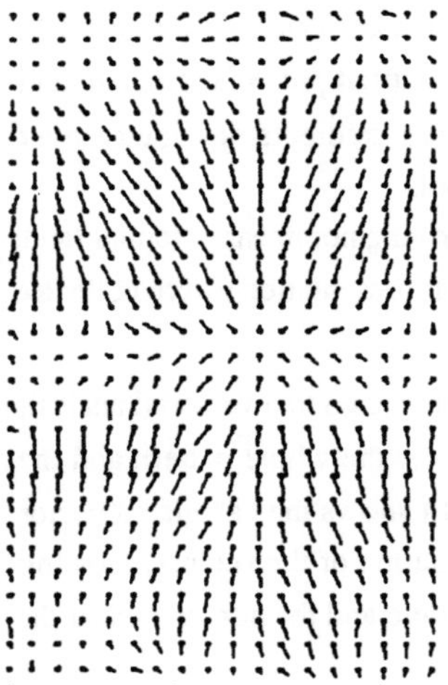

Fig.14: Calculated vector field of the averaged displacement vectors of 10 Ss

towards or away from the attractors. Again even subliminal verbal suggestions, e.g. "down" or "up", given to the subjects below the threshold of conscious recognition (masked by music) during the task of point reproduction has a significant effect (figure 18). Like in multistability the dynamics underlying the process of cognitive order formation are obviously open to the macro-

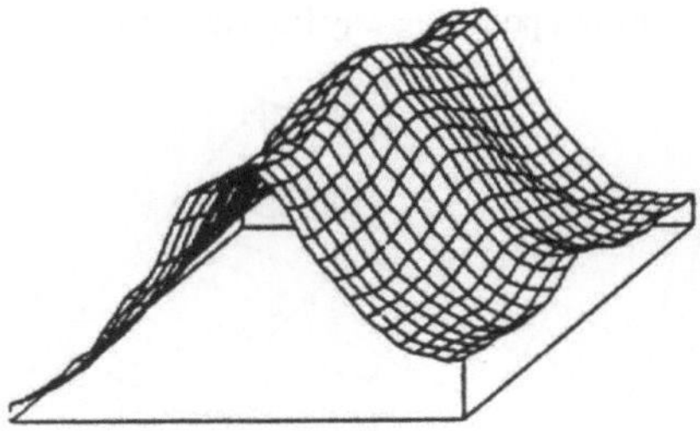

Fig.15: Gradient potential of the calculated vector fields

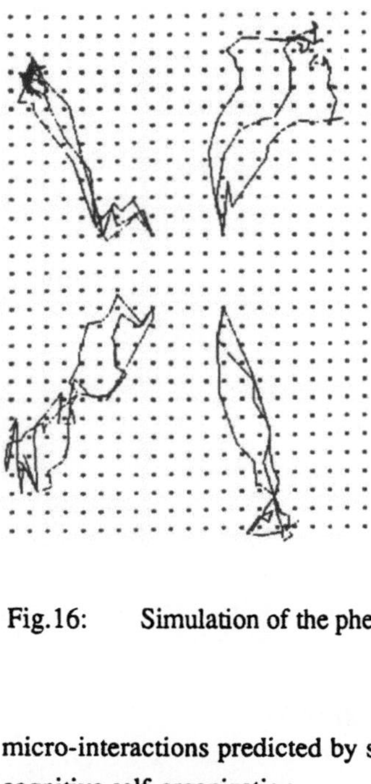

Fig.16: Simulation of the phenomenon of the wandering dot

micro-interactions predicted by synergetics. Meaning is an order parameter of the process of cognitive self-organization.

7 Attractors of complex behavior

In the last section the concept of attractors is used as an organizing metaphor to describe complex behavior. During the first months of life the satisfaction of the basic needs of a baby (sleeping, eating etc.) follows a chaotic attractor, i.e. the baby cries whenever it feels unwell. But after some weeks the babys behavior transites to a periodic atrractor provided that it is not restricted

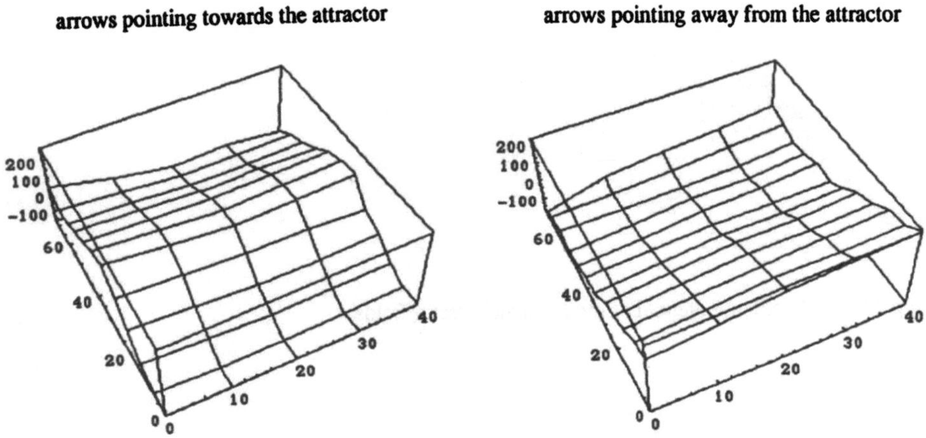

Fig.17: Changing the gradient potential by meaning

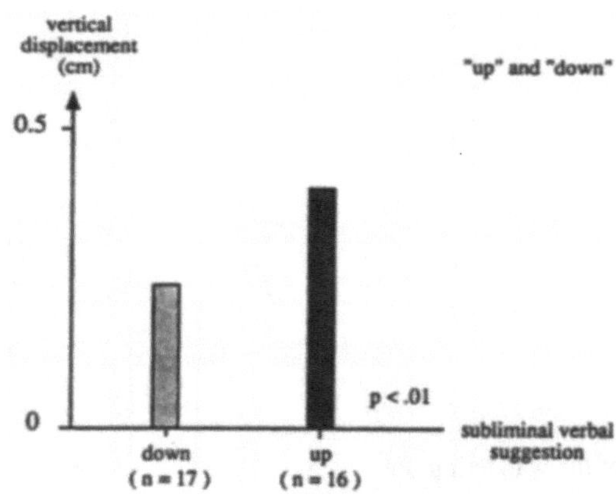

Fig.18: Changing the displacement vectors by subliminal suggestion

by the environment. If the baby may sleep and wake whenever it likes and gets food whenever it cries it will find a temporal organization of its need satisfaction by itself (see figure 19). This time schedule is very similar to the one the parents might have enforced with much more educational effort (26).

Growing up the child begins to organize its needs and goals in a spatio-temporal attractor landscape which represents its environmental field. In principal such a field can be assessed empirically. Imagine a potential field similar to those found in perception which model the dynamics of human actions. The potential landscape equals the sum of all needs, motivations,

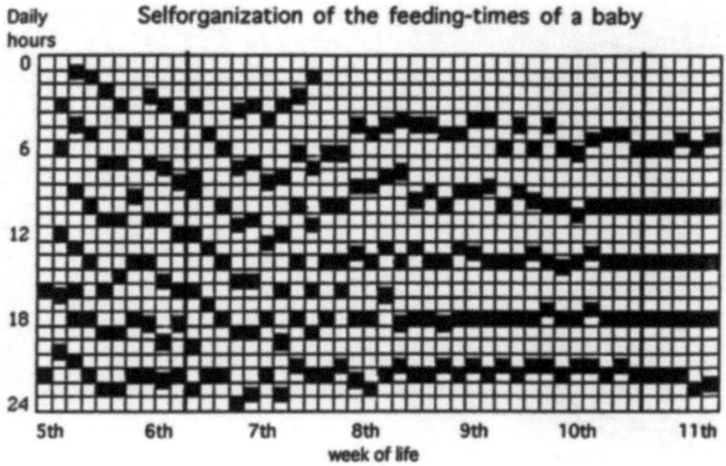

Fig. 19: Record of the feeding-times of a baby during the 5th to the 11th week of life (see 26)

evaluations, and goals which characterize a person at a given moment. In this view the persons locomotion follows attractor valleys and is repelled by hills. In such a way it should be possible to measure such an assumed landscape by a method similar to that explained above in the perceptual field. The task of the subject would be to guess the distances to all relevant points in his experienced environmental field. Using the displacement vectors between the physical and the guessed coordinates the potential field can be calculated. But the underlying psychological processes are presumably not stable enough and too easily influenced by the method of measurement itself. In the decribed way the measurement is probably only a "Gedankenexperiment" (thought-experiment).

In psychology there have always been ideas to differenciate in a similar manner between the psychological environment and the physical world. Kurt Lewin [27] made clear that the psychological field contains only a few areas which are free for locomotion. Others are closed by psychological or physical barriers. These barriers may be defined only by the motivation of the subject itself. The psychological fields are structured in certain ways. For instance a child which is educated very liberaly has many degrees of freedom for moving around in its environment while another child, grown up with many restrictions, may have a small space for own movements (see figure 20). Kurt Lewin called the valleys in the environmental potential field "positive valences (V+, Aufforderungscharakter)" and the hills are defined as "negative valences (V-)". Valences in Lewin's so called life space (the psychological field or environment) attrack and repell human behavior.

The following example for attractors in human action is a case study from the legal psychological practice [28]. Actually the deed to be explained in the attractor-concept is a crime

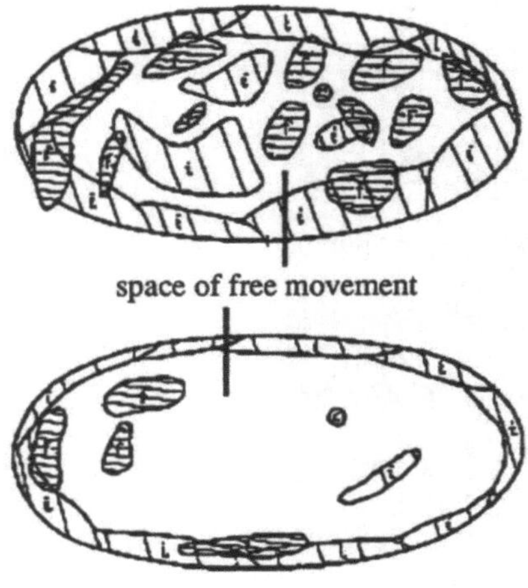

Fig.20: Psychological stucture of the environmental field

of passion. If the deed is to be explained as a consequence of the intrinsic dynamics the accused person is to be regarded from the viewpoint of diminished culpability.

The female offender (Alice) had been married to her husband (Bernhard) 12 years prior to the deed. She had been together with him for two years before the marriage. One year before the marriage their son was born. Alice had pressed for the marriage, since she worshipped Bernhard and saw him as the ideal partner for a happy family life. But soon she was disappointed by the marriage, since Bernhard repeatedly committed criminal offenses, worked irregularly and consumed a considerable amount of alcohol. He went on extended drinking tours two or three times per week. When Bernhard was drunk he was very aggressive. He treated Alice in a way which can be called sadistic. Already in the third year of their marriage Bernhard beat up Alice so strongly, that she finally became unconscious. In the further course of their marriage Bernhard began to threaten her with a knife and once injured her hand with it. In such situations he often threatened her: "I'll kill you, I'll stab you."

Alice held a regular job and was esteemed as a reliable and diligent worker. When Bernhard maltreated her, she tried to soften him with calming and soothing behavior. Several times she had to leave the apartment in order to escape from Bernhard. On these occasions she also took with her the daughter born in the fifth year of their marriage, whom she used to put to bed fully dressed and ready for flight. The firstborn son had been staying with Alice's mother since right after his birth. While the son had been repeatedly kicked by Bernhard, Alice had succeeded to protect the younger daughter. In spite of all this Alice never gave much thought to the possibility of divorce, because she continued to hope for a harmonious marriage and to long for a change in Bernhard's behavior. Bernhard promised again and again to start working regularly and to quit drinking. A complete marital break-down and alienation between Alice and Bernhard was averted by their fulfilling sexual relationship during the times when Bernhard was sober. When Alice thought about a divorce she used to come to the conclusion that this would mean a social decline for her, that Bernhard would still not leave her alone and that she did not want the children to loose their father. Therefore she tried to keep up the appearance of a harmonious marriage. When she had been beaten by Bernhard she went for a demonstrative walk with him arm in arm. She pretended that Bernhard held a regular job and covered bruises and grazes with make-up or sunglasses. In everyday life Alice behaved in a markedly reserved, sensible and conformist manner. She usually took the blame for the marital fights. Her rare defence reactions were limited to verbal statements such as "You're crazy!". Alice tried repeatedly to hide Bernhard's knife, but he always got himself a new one immediately.

On the evening before the deed Bernhard hit the eight-year-old daughter while inebriated. The girl fled to her mother and cried for help. Alice pleaded with him and asked him to hit her instead. Bernhard did this, but continued to hit the girl in-between. After Alice had brought the girl to bed Bernhard dumped the contents of the ashtray on the floor and forced her to pick up the cigarette stubs. On the following morning Alice drove Bernhard and the girl to her parents in law and went to work. Bernhard soon left the house and visited several bars. Around five o'clock in the afternoon he went back and started a fight with his father, who in the course of the fight injured him with scratches. When Alice came back around six o'clock and heard about the fight she immediately wanted to drive home with her family. Bernhard forced her however to visit several more bars with him beforehand. Shortly before eight o'clock Bernhard agreed to drive home. During the drive he talked about the fight with his father and remarked that he would have stabbed him a long time ago if he had not been his father.

After arriving in the apartement Alice took a short break in the living room. She was exhausted, since she had hardly slept the night before, because she was distressed about Bernhard's treatment of the daughter. She had worked all day and had been forced against her will to the drinking tour with Bernhard. Furthermore she was in the premenstrual phase and as always in this phase felt more nervous and fearful than usually. Bernhard set down across from her, took out his knife and put it on the table. Alice left the room and went into the kitchen to prepare herself a sandwich. There she heard how Bernhard told her to again accompany him to a bar. Alice answered him that she would not do that and took hold of a breadknife, which she then laid down on the kitchen table. Suddenly Bernhard stood next to her and shouted at her: "What, you're not doing what I'm telling you? - Then I'll kill you today and stab you!". In that moment Alice felt the edge of his knife on her neck. She turned around and said: "I think you're mad. Take your knife away!". Although she was scared, at that moment Alice did not think of a disaster. Not until Bernhard approached threatening her with the knife did she feel fear of death. She tried to fence Bernhard off and pushed him against the kitchen table, which shook so that the breadknife fell down from the table and onto the floor. Bernhard then cornered Alice in such a way that she could not escape. Alice pushed him again back in the direction of the table, so that the table was also moved. Visibly in rage Bernhard again approached Alice and tried to grab her. With this he slid on the floor and fell down backwards. Alice reached for him and tried to keep him from falling because she was afraid Bernhard would blame her for his fall. They both went down. Alice sat half kneeling next to him and he lay with his upper body upright in her arm. When in this position Bernhard also threatened to kill her, Alice happened to see the breadknife on the floor. She lifted it up with her right hand and at eight thirty she stabbed Bernhard repeatedly with the knife, which penetrated his thorax nine times. Afterwards she sat for a while in the kitchen and said to Bernhard: "I loved you. It was all your fault."

In that case the woman had during all her marriage only one attractor of behavior. It is well known from phylogenetic studies of animal-ethology that there are three different possible ways of behavior in such a situation of permanent threat: escape, apathy (pretending to be dead) and attack. Alice showed mostly avoidance-behavior (escape). Sometimes she also was apathetic. But she never followed the strategy of attack or aggression against her husband. Evidently she had habituated to this avoidance-behavior because she always hoped for a harmonious marriage. Yet, the day before the deed, she experienced that her husband can be attacked and injured and this was the first time that such a possibility came into her mind as a new attractor of behavior. Furthermore, the very complex social situation of that family was destabilized by the fact that for the first time the husband had beaten her small daughter. Brood care-behavior was actualized in Alice's behavioral inventory, which is usually related to attack behavior if an aggressor comes into sight.

Additionally to the situational destabilization and the emergence of the new attack-attractor there was an extraordinary destabilization in Alice's psychophysical state: She was extremely exhausted, returning from her work, she had much lack of sleep, she had drunk alcohol during the bar visits, and she was in the premenstrual phase. In the critical situation she was at the bifurcation point between the old escape attractor and the newly established attractor of attack. The knife lying on the floor resulted in falling into the new attractor. A small cause had an enormous effect: she killed her husband. Before that she would never have been able even to think about such a deed. The killing action was in a way lawful but the situation in which it was performed was not predictable. Many conditions worked together to create an instable situation. At the same time a new action attractor had emerged.

In the traditional context of explanation such a crime of passion cannot be explained at all. Concepts like "inconsistency of behavioral style" or "strangeness to the personality" are used to describe the fact that there seems to be no possibility to conceptualize such abrupt behavioral changes. As we have shown in examples of different complexity nonlinearities are of special relevance for understanding cognitive processes. The theoretical concept of synergetics is a tool which enables the integration of these phenomena in psychological research.

Acknowledgements

This research work has been supported by the Deutsche Forschungsgemeinschaft (DFG). The authors wish to thank Daniel Strüber and Dr. Michael Kobs for their assistence and Prof. Dr. Günter Vetter for calling attention to the second example of a phase transition in the learning chapter.

References

(1) H. Haken & A. Wunderlin: Die Selbststrukturierung der Materie (Vieweg, Braunschweig 1991).

(2) E. Basar, H. Flohr, H. Haken & A.J. Mandell (Eds.): Synergetics of the Brain (Springer, Berlin 1983).
H. Haken & M. Stadler (Eds.): Synergetics of Cognition (Springer, Berlin 1990).

(3) W. Köhler: Jahresberichte für die gesamte Physiologie und experimentelle Pharmakologie 3, 512-539 (1925).

(4) M. Stadler & P. Kruse: Gestalt Theory 8, 75-98 (1986).
M. Stadler & P. Kruse: The Self-Organization Perspective in Cognition Research: Historical Remarks and New Experimental Approaches, in: Synergetics of Cognition, ed. by H. Haken & M. Stadler (Springer, Berlin 1990).
P. Kruse, G. Roth & M. Stadler: Gestalt Theory 9, 150-167 (1987).

(5) K. Holzkamp: Psychologische Rundschau 21, 1-22 (1970).

(6) G. Johansson: Configurations in Event Perception (Uppsala 1950).

(7) H. Haken: Synergetics (Springer, Berlin 1977).

(8) M. Stadler & P. Kruse: Philosophical Psychology 6 (to be published 1993).

(9) H. Haken: Pattern Formation and Pattern Recognition - An Attempt at a Synthesis, in: Pattern Formation by Dynamic Systems and Pattern Recognition, ed. by H. Haken (Springer, Berlin 1979).

(10) H. Haken, J.A.S. Kelso & H. Bunz: Biological Cybernetics 51, 347-356 (1985).

(11) M. Stadler, P.H. Richter, S. Pfaff & P. Kruse: Psychological Research 53, 102-112 (1991).

(12) D.L. Bryan & N. Harter: Psychological Review 4, 27-53 (1897).
D.L. Bryan & N. Harter: Psychological Review 6, 345-375 (1899).

(13) D.V. Cross & H.L. Lane: Journal of the Experimental Analysis of Behavior 5, 487-496 (1962).

(14) P. Kruse: Delfin 6, 35-57 (1988).

(15) T. Ditzinger & H. Haken: Biological Cybernetics 61, 279-287 (1989).
T. Ditzinger & H. Haken: Biological Cybernetics 63, 453-456 (1990).

(16) P. Kruse, M. Stadler. & D. Strüber.: Psychological Modification and Synergetic Modelling of Perceptual Oscillations. In: Rhythms in Physiological Systems. ed. by H. Haken & H.P. Koepchen (Springer, Berlin, 1991).

(17) see note 16.

(18) C.-F.Graumann: Zeitschrift für experimentelle und angewandte Psychologie 3, 605-661 (1956).

(19) P. Kruse: Gestalt Theory 8, 141-143 (1986).

(20) C. Basar-Eroglu, D. Strüber, M. Stadler, P. Kruse & E. Basar: Slow positive Potentials in the EEG during Multistable Visual Perception (submitted).

(21) F.C. Bartlett: Remembering (University Press, Cambridge 1932).

(22) B. Mandelbrot: The Fractal Geometry of Nature (Freeman, New York 1970).
H.-O. Peitgen & P.H. Richter: The Beauty of Fractals (Springer, Berlin 1986).

(23) F. C. Bartlett: The Mind at Work and Play (Allen & Unwin, London 1951).

(24) J.J. Gibson: The Perception of the Visual World (Houghton Mifflin, Boston 1950).

(25) see note 11.

(26) W. Metzger: Schöpferische Freiheit (Kramer, Frankfurt/M. 1962)
W. Metzger: Psychologie in der Erziehung (Kamp, Bochum 1976).

(27) K. Lewin: Principles of Topological Psychology (McGraw-Hill, New York 1936).

(28) T. Fabian & M. Stadler: Applying Chaos Theory to Delinquent Behavior in Psychosocial Stress Situations, in Psychology and Law, ed. by F. Lösel, D. Bender & T. Bliesener (De Gruyter, Berlin 1992).

Pattern Formation in Complex Cognitive Processes

J. Kriz

Fachbereich Psychologie, Fach: Klinische Psychologie,
Universität Osnabrück, D-4500 Osnabrück, Fed. Rep. of Germany

1 Introduction

Today, the so-called "systems approach" has become one of the most important fields in clinical psychology. Especially the practice of psychotherapy with families is very famous and, as a consequence, new theoretical concepts have emerged in order to understand psychopathological and psychotherapeutic processes.

When we browse through the literature on systemic clinical psychology we find that most authors refer to interaction patterns. This fact represents a valuable change of perspective: mental illnesses that were considered "individual" in other approaches (whether they were caused by inner conflicts, by learning, by "faulty thinking", or whatever) are analyzed in terms of their status and their function for the communicative structures in a social system–especially in the family. A "pathological" family thus regulates itself through patterns of communication that reflect the type of symptom. It is evident that these patterns of interaction are the results of self-organized processes. There are no external orders or rules given in a way that each person has to act in a well-defined way to fulfill these, but the persons act together by some kind of mutual understanding, each one acting so as to fulfill the rules of a game.

In spite of the deepening interest of clinical psychologists in family therapy and related concepts, it seems that the discussion suffers from an inability to account for the manner in which patterns of interaction emerge and become transformed. We may insist that the environmental context of these family interactions must be thoroughly understood and analyzed before an adequate account of communication structure and their transition is possible. On the one hand, attention must be drawn in a dynamic-systemic circularity to the fact that patterns of processes in systems have only been developed relative to the patterns of processes in the meta- systems (the "environment"), and now both perpetuates the pattern of interactions as well as being perpetuated itself by these interactions. Accordingly, it would be difficult to overestimate the value of processes and their rules of lager social systems for the interaction patterns of a family (e.g. the economic situation, social values and norms etc.).

On the other hand, by focusing on family interactions one may become blind for the fact that the persons who have come together (usually the spouses) have brought certain experiences and habits into this relationship, which have an effect on their setting up certain rules. "Patterns of the interactions", therefore, also reflect expectations, interpretations, definitions of "reality", etc. by the family members. Moreover, communication, something interpersonal, a mark of the system "family", is directly related to intrapersonal aspects (e.g. biochemical and neuronal processes, muscular and general physical constitutions, experiences and the ways to store them, self-esteem, etc.).

Another important point to be made about systemic therapy is the fact that not only patterns of interaction are the results of self-organized processes but, by the same logic, thoughts, cognitions, processes of the central nervous system and, on the other hand, society and several sub-systems of the society can be understood as self- organized processes, too.

As a consequence, we find a hierarchy of phenomena, each level may be described in terms of self-organized processes. The question to be answered therefore becomes: "How can we conceptualize these multilevel interactions of micro- and macroscopic processes (from any given standpoint) in terms of self-organization ?" In contrast to other conceptions of self-organization (e.g. the concept "autopoiesis" by Maturana & Varela 1987, or Luhmann 1984), the so-called "synergetics" by Haken (1981) gives a profound answer to the question of how we can conceptualize the emergence of self-organized patterns and which are the principles by which this order may change to a new pattern. Moreover, Haken's synergetics stresses the relationship of systems and subsystems by "order parameters" (or macroscopic field-variables) which evolve out of a complex multicomponent systems on a microscopic level through self-organization, and by "control parameters" which represent the system's environment in a rather unspecific manner — but the change of them can induce a qualitative and specific change in the behaviour of the system.

According to this interdisciplinary field of synergetics, I have proposed to simulate family processes with the help of equations taken from population dynamics (Kriz 1990, 1991, 1992). In these studies, the system has been conceptualized on the level of interactions (or, more precisely, categories of interactions). However, it has been emphasized that the control parameters do not only reflect influences from other (categories of) interaction(s) but also reflect the system's "environment" in a special manner: one component of the control parameters represents the influence of "internal" (intrapersonal) processes, while another component of the control parameters represents the influence from meta-systems of the family (e.g. society — or, better, parts of society). Therefore, the control parameter in each equation has been decomposed into three components which allow one to model the relationships not only to processes on the same level of the system hierarchy (here: "communications") but also relationships between processes in the family and in "society" on the one hand, and between processes in the family and in the individual person (e.g. especially cognitive processes — or, better, "affective-cognitive" processes, see Ciompi 1988) on the other hand (Kriz 1992, p.160). As a consequence, according to Haken, an order parameter at one hierarchical system-level acts can be understood as control parameter at another one.

Moreover, Haken has pointed out that we may expect the same relationship between affective-cognitive processes and neuronal processes — which is of strong interest, because it concerns the mind-body-problem: "behavioural patterns, percepts, thoughts, and other mental processes can be represented by order parameters (or sequences of them). They describe the system at the macroscopic level and they are the means by which, for instance, we communicate with each other. At the same time they prescribe the order of the micro system, e.g. the neurones, which in turn determine the macroscopic order parameters. As I have stressed time and again, a circular causality is present, where the order parameters describe the behaviour of the individual subsystem which in turn determine the order parameters" (Haken 1990, p.11).

These considerations suggest that it may be of value to take into account the pattern formation (and pattern transition) in "affective-cognitive" processes, even if one wants to explain interactions between persons. The purpose of this study is to present some empirical work which has been done in order to investigate this question.

2 Pattern Formation and Cognition

The concept of "patterns" or "structures" in cognitive processes is not at all new. For example, the Swiss biologist and psychologist Jean Piaget has formulated the theory of cognitive "schemata" which emerge and change through "assimilatory" and "accommodatory" interaction with reality (Piaget 1952). Moreover, the psychologist George A. Kelly has published "the psychology of personal constructs" (1955). This book is of strong interest, because it stresses the aspect of cognitive dynamics and "understanding" as constructing patterns in the process of reality.However, even some decades earlier, concepts of self-organization and pattern formation in cognitive processes has been put forward by the research program of Gestalt psychologist. Gestalt psychology founded by Wertheimer, Koffka and Köhler has been of great importance in the twenties and thirties of this century. During those two decades, dozens of studies in Gestalt psychology have evaluated a mass of prominent features of order formation in perception and cognition — for example, the concept of "Prägnanz", bi- and multistability of perception, or invariance properties of "Gestalten".

Although the research program of Gestalt psychology was introduced nearly 80 years ago, Stadler & Kruse (1990) as well as Haken stressed the fact that there are astonishing correspondences between the terminology and the argumentation of Gestalt psychology and modern self-organization concepts — especially synergetics: "Throughout the whole history of Gestalt theory the autonomy of order formation in cognition was the explicit starting point of conceptualization and experimental work. Therefore Gestalt theory offers a lot of well- described phenomena and stimulating ideas for the elaboration of self-organization theory of cognition" (Kruse et.al. 1992, p.103).

As a consequence, Stadler and Kruse carried the early Gestalt psychologist's work further, paying special attention to phenomena which are important in the context of synergetics. They rediscovered a procedure which has been invented by Sir Frederic Bartlett in the thirties (Bartlett 1932). The "Bartlett-scenario" (as they called it) uses recursive iteration procedures as a methodological tool for the demonstration and analysis of order formation. The basic idea is to ask a person to reproduce a given visual or cognitive pattern (for example, a short story) and then to use this reproduction as a stimulus for the next reproduction (by another or by the same person) and so on. In everyday life this procedure is well known in the field of rumour generation. Such serial reproductions may be regarded as data for a detailed analysis of slight systemic tendencies in the self- organization process in perceptive and cognitive systems. While Bartlett's methodology has been used, for example, in the linguistic analysis of text understanding and remembering short stories, Stadler & Kruse (1990) reported their recent perceptual studies in which they showed that serial reproductions of random dot patterns tend to highly ordered structures which are steady states — or, in the terminology of systems dynamics, attractors — because they are no more subject to alterations in further reproductions (see figure 1).

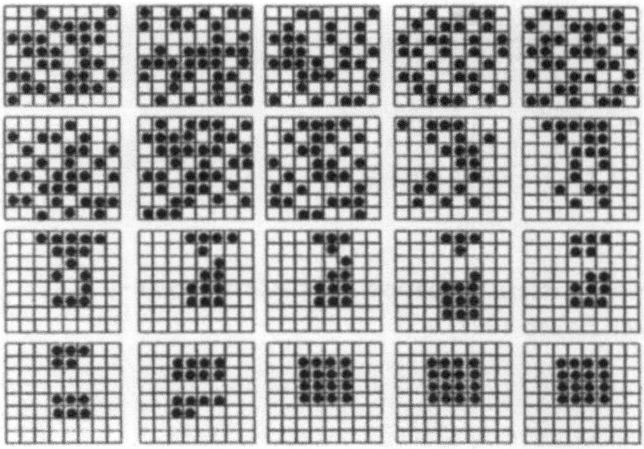

Fig.1: Serial reproduction of dot patterns (to be read line by line); upper left: initial stimulus pattern (After Stadler & Kruse 1990).

In our studies (Runde, Kessler & Kriz 1992, Kriz, Kessler & Runde 1992, Kriz & Kriz 1992), we also used the Bartlett-scenario. However, this procedure was not applied to analyze visual processes but to investigate cognitive processes in the domain of concept formation, attribution, attitude, stereotype, and prejudice — in short: the formation of order in the cognition of our social world. Attribution theories "draw on the human thought processes to explain why we do what we do...Attribution theorists point out the way we selectively process information and the way we change our thinking to match our behaviour, and vice versa. (We reject or forget information that's inconsistent with our beliefs...)" (Papalia & Olds 1986).

We assume that these cognitive processes have to do with pattern formation and pattern recognition which have been studied by the Haken and his group with respect to the synergetic computer. Their findings suggest that pattern recognition is nothing but pattern formation. This is demonstrated in figure 2 (schematically) and figure 3.

When part of the subsystems are in ordered state, they generate their order parameter which then enslaves the rest of the system and produces a totally ordered pattern (fig.2, left); on the other hand, when some features

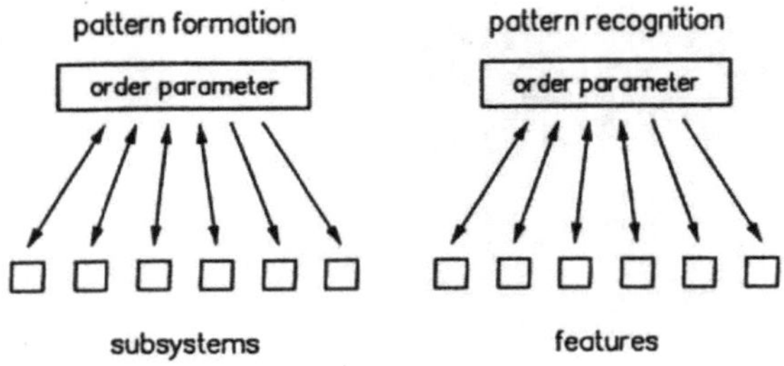

Fig.2: The analogy between pattern formation and pattern recognition; compare text (After Haken 1992).

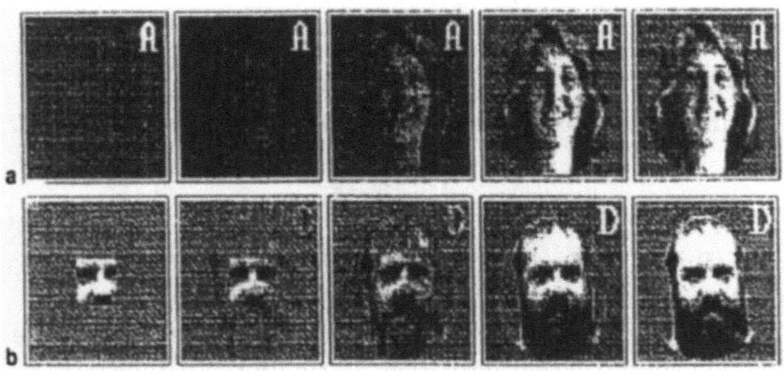

Fig.3: Visualization of the pattern recognition procedure. The recognition: synergetic computer recognizes a face and its name (a letter) out of an initially given set of features — (a): a strong superposition of noise on it; (b) just a small part of it (After Haken 1992).

of a pattern are given, they generate their corresponding order parameters which then restores the whole recognition pattern (fig.2, right).

Of course, the synergetic concept of pattern formation and pattern recognition is nice in the case of a synergetic computer where noisy or distorted patterns have to become recognized. In contrast, if this is a model for what happens in associative memory — and there is a lot of evidence for it — then parts of a set of data is "completed" in a unique fashion. However, this might not only happen when we "recognize" the face of an old friend from a short visual impression, or a word from some letters, or the context of a situation from some pieces of information. By the same logic, we might generate and construct patterns on the basis of very little or unclear information while we believe we restore or recognize the information. It is quite likely that this happens in cases of prejudice, stereotypes etc.

This is very relevant to clinical psychology, too. In the "person-centered system approach" (Kriz 1985, 1991) I have stated that patterns or rules in the communication process of a family is related to categories of perceptive and cognitive processing of information (e.g. communications of other family members) — especially in families where observers describe a "lack of flexibility" and a "rigid" behaviour. Therapist often hear the argument of single family members: "Why should I change my behaviour — the others wouldn't notice it". This seems to demonstrate that only a few pieces of information are taken and may already suffice as the basis for pattern formation or "recognition" — that means that the order parameters of the cognitive patterns (=interpretation of the behaviour) follow a trajectory towards an attractor, which might be of low dimensionality and, moreover, there may exist only very few attractors. If a family member experiences only very few categories (attractors) of behaviour, motivations etc. there is no reason to "react" to these in a high differentiated manner. As a consequence, he or she will practice a small repertoire of communication behaviour. As in the whole communication process of a family the communication behaviour of one person is closely related to the perception-, cognition-, and interpretation-process of an other one — and vice versa — this dynamic may show a decreasing repertoire both in "expressions" and "impressions" of communication behaviour. Again, the arguments appear to indicate that it is important to look at attractors in the dynamic of cognitive processes.

3 Experiments

In order to investigate the dynamic of pattern formation and/or recognition in cognitive processes, we used the Bartlett-scenario, as mentioned above.

In our first study (Runde, Kessler & Kriz 1992) we were interested in finding out whether a process of associative response, ruled by the Bartlett-scenario, have attractive features (we avoid to say: "have an attractor", because the term "attractor" has a strong mathematical notion). In this experiment, a Subject is asked to respond to a given word with a word that comes into mind. 450 Words were given to one subject — however, only the first 30 words were selected from a personality inventory, while words 31-60 were a random sequence of the subject's responds, words 61-90 were

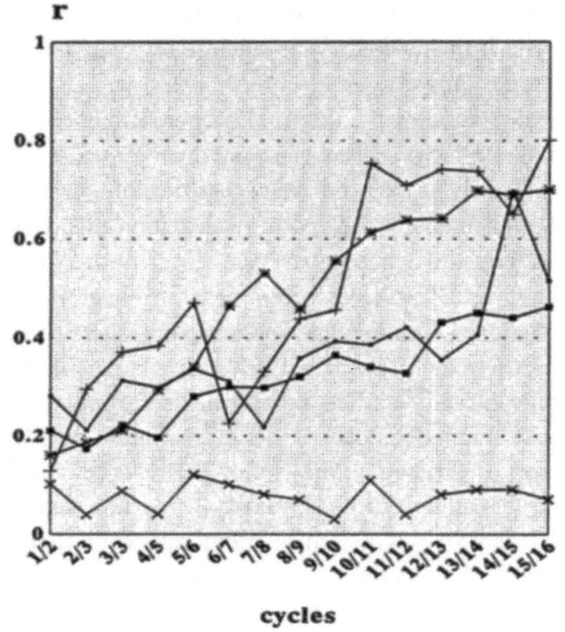

Fig.4: Cophenic correlation coefficients in a serial (re)production association experiment (see text).

a random sequence of the responds of the responds, and so on. This can be understood as a serial (re)production procedure with 15 cycles.

When a subject had finished this part of the experiment he or she were asked to classify the 30 words of each cycle with respect to their resemblance (of course the sequence of cycles in the sequence of words in each cycle has been matched by random). No further instruction — for example, concerning the number of classes or any content — was given.Figure 4 shows the cophenic correlation coefficients (Hubert and Baker 1977) between two successive cycles for 4 subjects. If there were no attractive dynamic, the correlations should be in range of random or at least vary in a random manner. In contrast, the correlation coefficients increase in a systematic manner for all 4 subject and become rather high (.44 until .80).

To get some evidence that the increasing resemblance is not an artefact of the formal procedure, a fifth subject was asked to classify the responses of one subject in the same manner as the subject did with the material produced by themselves. Again cophenic correlation coefficients were computed — but now they varied in the random range (curve between .02 and .12 in fig. 4).

In a second experiment (Kriz, Kessler & Runde 1992) subjects were given 10 statements of a German personality inventory (FPI-R). The items of this inventory are described by 10 factors of classical factor analysis and each of the 10 statements was chosen from exactly one dimension. To give an example, one statement was: "I am often in situations with lot of stress", another: "sometimes I feel gloating".

The subjects were told to assume that these ten items are statements made by "Mr. K. from W." about himself when they had "met him during their holiday in Spain". They had enough time to read the statements. The subjects then got a questionnaire with 200 more items from that personality test. They were asked to guess whether "Mr. K. from W." would have agreed to that statement about him. Additionally, they were asked to scale how sure they are in their judgement on a rating scale between 1 and 5. These 200 items were presented on 20 pages of paper, 10 items/page. Moreover, each page with 10 items represented a random sequence of exactly the 10 personality dimensions. As a consequence, this procedure can again be understood as a serial (re)production procedure with 20 cycles, 10 items each.

In order to guess the answer of Mr. K., subjects have to take 200 decisions. For each decision it is necessary to produce a notion of Mr. K.'s attitudes and personality. On the other hand, the real information about Mr. K. was one statement on each personality dimension — i.e. a rather equal distribution in a complex semantic space. Talking about pattern formation in a semantic space rises the question how to describe a "pattern". In the reported experiments of Stadler & Kruse it was clear to take the two dimensions of the plane. But how to describe semantic patterns?

To keep things simple, we focused on the personality factor analysis dimensions. These, of course, are not the dimension of the semantic space one probably would like to know in order to describe an multidimensional attractor landscape. However, if the personality test is not worthless at all, the personality dimensions of that test should reflect something what clinical psychologists would find relevant in order to describe phenomena in that area. For that reason, we were interested to find out whether the process of 200 guesses (20 cycles) leads to stable judgements in some personality dimensions.

Figure 5 illustrates the results of one person. If the subject constructs a clear notion (in a specific dimension), the guesses change from randomly "yes" or "no" to a stable guessing. Recall that subjects did not get any explicit information about "dimensions" or about what statements belong to what dimension. Moreover, in figure 5 we turned the direction of guesses in the way that points under the line represent "disagreement" with the concept

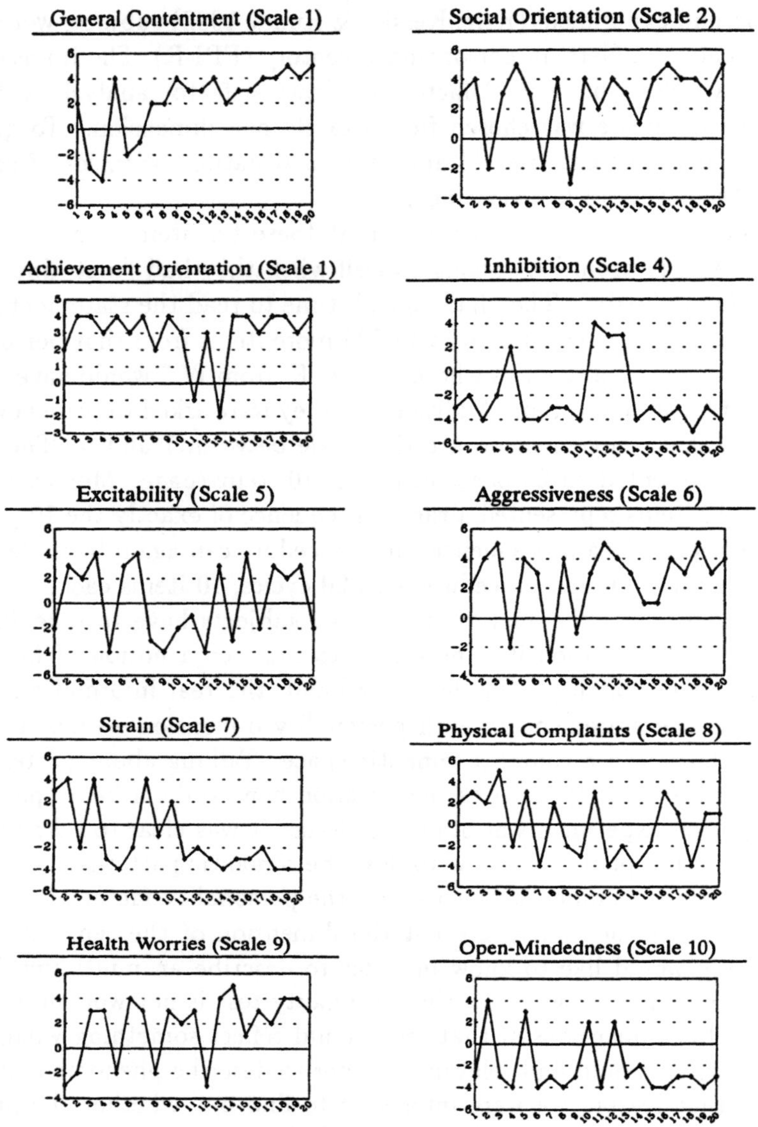

Fig.5: Serial guesses by one typical subject. Guesses are split into 10 personality scales (compare text).

	"normal" persons					"schizophrenic" persons				
	Vp 9	Vp 10	Vp 11	Vp 12	Vp 13	Kl 1	Kl 2	Kl 3	Kl 4	Kl 5
Scale 1	*	*		*/0	*					
Scale 2	*	*/0	*/0	*/0	*	*				
Scale 3	*	*	*/0							
Scale 4	*	*			*					
Scale 5			*/0	*/0						
Scale 6	*	*/0	*/0	*/0	*					
Scale 7	*	*	*/0	*	*					
Scale 8		*	*/0	*/0						
Scale 9	*	*	*		*					
Scale 10	*		*							

Fig.6:(compare text)

of that personality factor while points above the line represent "agreement" — however, to be in "agreement" with a personality factor, subjects had to judge about half the time "yes" and half the time "no" (in a random sequence). Consequently, to guess "stable" in a specific dimension means that the subject has to construct a clear notion of that personality dimension. This happened, for example, at "scale" 1, 6 or 7 in figure 5. In contrast, for "scale" 5 and 8 no clear notion turned out (in these 20 sequences): the guesses vary randomly — or, in other words, in this personality dimensions no attractive process is found for that person.

A third characteristic sequence of guessing may be distinguished: the guesses are from the very beginning "stable" at a specific scale. This is not the case for the subject's data in fig.5 but was observed in the data of other subjects (described later). One more interesting result of that study is the fact, that all "normal" (see later) subjects show attractive guesses only in some dimensions while in other dimensions random guesses indicates that no notion of Mr. K. has been formed. Moreover, these dimension vary from subject to subject (we had 10 subjects - 5 in a pretest with only 15 cycles and 5 with 20 cycles).

In contrast, 5 "clinical" subjects — patients of a psychiatric hospital with the diagnosis "schizophrenic disease" but not in an acute phase or attack of symptoms - did not show any stable guessing at the personality scales. They were clearly unable to form a pattern of the personality of Mr. K. (described at these 10 Dimensions). Figure 6 illustrates this contrast between 5 "normal" subjects of the main study and 5 "schizophrenic" subjects. A

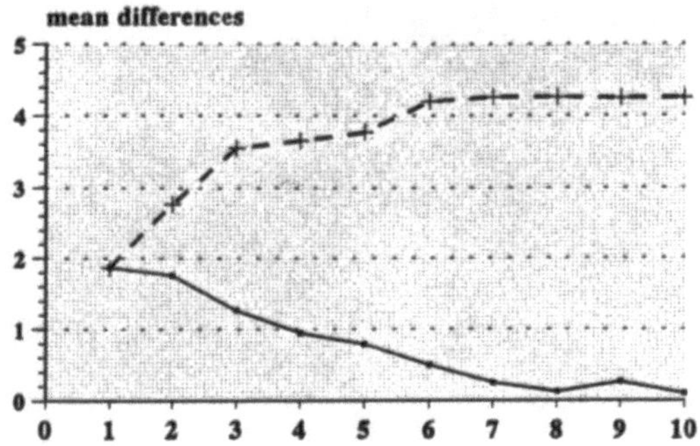

Fig.7: Dynamics of mean differences between two successive "recalls" of attributes related to a fictive person (solid line) and between these "recalls" and the initial stimulus pattern (initial description of attributes) (dotted line). Data from one typical subject.

"*" indicates that at least the last 5 cycles are guesses in the same direction of the certain factor while "0" stands for the above discussed case in which the subject had stable guesses just from the beginning (the first 5 guesses were in the same direction, too).

A third experiment demonstrates the independence of pattern formation from given patterns of "reality": In this study (Kriz & Kriz 1992) 30 subjects were given 72 statements which describe a fictive person. These statements were very short because the just combined 72 pairs of adjectives ("intelligent-unintelligent", "happy- unhappy", "friendly-unfriendly", etc.) with 4 qualitative categories ("extreme", "very", "rather" "little"). Accordingly, the statements were like: Person X is "very intelligent", "extreme unhappy", "rather friendly" etc. A subject had to read these 72 statements in as much time as he wanted. After some minutes of doing mental arithmetic (in order to have some other cognitive activity) he was asked to reproduce the statements by making crosses in a list with 72 pairs of adjectives — each pair in one line with a nine — point-scale between (2 x 4 categories plus "I don't know").

These answers were entered into a computer while the subject again was doing some mental arithmetic. The next step then was to present the subject a printout of his answers — however, transformed again in statements and printed out by the computer. This procedure was repeated 10 times.

In figure 7 the ordinate represents a measure of change from one cycle to the next one (which is just the mean of all differences) while the abscissa shows the 10 cycles of serial reproduction ("0" is the initial material). The solid line is computed from the differences between "input" and "output" in each cycle. Accordingly, a difference of 0(or nearly 0) stands for the fact, that the description of the person is perfect (or nearly perfect) repeated. In fig.7 this is the case after 7 cycles while at the first cycles we find rather big changes (a maximum change would be at about 6 because from one extreme "−4" to the other "+4" is the difference 8, but if we start from "−1" the maximum possible difference is 5).

From the foregoing facts it is evident that the process of serial constructed descriptions of the person has an attractive dynamic. It is important to note that fig. 7 is typical to 25 out of 30 subjects. However, for 5 subjects no attractive dynamic can be seen (in 10 cycles): the differences between successive cycles - or, which is the same: between "input" and "output" - remain rather large.

Another important point to be made about the attractive dynamic (by these 25 subjects) is the fact that the descriptions of person's attitudes became rather quick stable, but these stable "descriptions" turn out to be merely constructions: The dotted line in fig. 7 represents the mean differences of each cycle to the initial description. It could not be more dissimilar!

A last finding from this study may be mentioned here. 3 different conditions were given to the 30 subjects (each condition to 10 subjects): the initial description was (i) rather consistent (ii) medium consistent and (iii) rather inconsistent (for example, the person was described as "very friendly" but later as "rather gruff"). While differences between successive cycles became rather fast 0 (or nearly 0) in all 3 conditions (for 25 subjects) the differences in the first two cycles were significantly higher for the inconsistent description than for the medium consistent description and were significantly lowest for the consistent description. Of course, the total difference of the stable pattern from the initial pattern follows shows the same effects.In conclusion it may be stated that constructive pattern formation/recognition changes an inconsistent "reality" more than an consistent "reality". This is rather likely because we all want to live in a consistent cognitive world rather than in a inconsistent one.

References

BARTLETT, F.C. (1932). *Remembering.* Cambridge: Cambridge Univ. Press.

CIOMPI, L. (1988). *The Psyche and Schizophrenia. The Bond Between Affect and Logic.* Cambridge, Mass: Harvard Univ. Press.

HAKEN, H. (1981). *Synergetics. An introducion.* Berlin: Springer

HAKEN, H. (1990). Synergetics as a Tool for the Conceptualization and Mathematization of Cognition and Behaviour — How far can we go? In: Haken & Stadler: *Synergetics of cognition*, pp. 2-31.

HAKEN, H. (1992). Synergetics in Psychology. In: Tschacher, Schiepek & Brunner, pp. 32-54

HAKEN, H. & STADLER, M. (eds.) (1990). *Synergetics of Cognition.* Berlin: Springer

HUBERT, L. & BAKER, F. (1977). Evaluating object set partitions free sort analysis. *J.math.psychol.,* **16**, 233-253.

KELLY G. A. (1955). *The psychology of personal constructs.* New York: W.W. Norton & Co.

KRIZ, J. (1985). *Grundkonzepte der Psychotherapie. Eine Einführung.* München: Urban & Schwarzenberg (3.Aufl.1989: Weinheim: Psychologie Verlags Union).

KRIZ, J. (1989). *Synergetik in der Klinischen Psychologie.* Forsch.ber. Nr 73, FB Pychologie, Osnabrück.

KRIZ, J. (1990). Synergetics in clinical psychology. In: Haken & Stadler, pp. 393-404

KRIZ, J. (1991). Mental Health: Its Conception in Systems Theory. An Outline of the Person — Centered System Approach. In: Pelaez, M.J. (Ed.): *Comparative Sociology of Familiy, Health & Education, vol. XX, (Barcelona, Espania)*, pp. 6061-6083

KRIZ, J. (1992a). Simulating Clinical Processes by Population Dynamics. In: Tschacher, Schiepek & Brunner: pp. 150-162

KRIZ, J. (1992b). *Chaos und Struktur. Systemtheorie Bd.1.* München: Quintessenz.

KRIZ, J., KESSLER, T. & RUNDE, B. (1992). *Dynamische Muster in der Fremdwahrnehmung.* Forsch.ber. Nr 87, FB Pychologie, Osnabrück.

KRIZ, W. C. & KRIZ, J. (1992). *Attrahierende Prozesse bei der Personen-Wahrnehmung.* Forsch.ber. Nr 88, FB Psychologie, Osnabrück.

KRUSE, P. ET.AL. (1992). Instability and Cognitive Order Formation: Self-Organization Principles, Psychological Experiments and Psychotherapeutic Intervention. In: Tschacher, Schiepek & Brunner, pp. 102-117.

LUHMANN, N. (1984). *Soziale Systeme. Grundrißeiner allgemeinen Theorie.* Frankfurt: Suhrkamp

MATURANA, H.R. & VARELA F.J. (1987). *Autopoiesis and Cognition. The Realization of the Living.* Boston: Reidel

PAPALIA, D.E. & OLDS, S.W. (1986). *Psychology.* New York: McGraw-Hill.

PIAGET, J. (1952). *The origins of intelligence in children.* New York: Internat. Univ. Press..

RUNDE, B., KESSLER, T. & KRIZ, J. (1992). *Serielle Assoziation — ein synergetischer Prozeß.* Forsch.ber. Nr 86, FB Pychologie, Osnabrück

STADLER, M. & KRUSE, P. (1990). *The Self-Organisation Perspective in Cognition Research: Historical Remarks and New Experimental Approaches.* In: H. Haken & M. Stadler (eds.): *Synergetics of cognition.* Berlin: Springer, 32-52.

TSCHACHER, W., SCHIEPEK, G. & BRUNNER, E.J. (1992). *Self-Organization and Clinical Psychology.* Berlin: Springer.

Modelling Pattern Formation in Ecological Systems

C. Wissel and F. Jeltsch

UFZ Umweltforschungszentrum Leipzig-Halle GmbH, Sektion Ökosystemanalyse, Permoserstr. 15, O-7050 Leipzig, Fed. Rep. of Germany

1. Introduction

Spatial inhomogenity is a typical characteristic of most ecosystems. This property is believed to be very important. It is a main factor which guarantees a high diversity of species. To some extent the spatial patterns of ecosystems are consequences of the spatial heterogeneity of the physical environment (geomorphology, soil humidity, exposition to solar radiation etc.). But there are numerous examples which demonstrate that spatial patterns in ecology can appear which are not caused by these physical factors. In these cases biological interactions in the ecosystems create the patterns in a synergistic manner. In addition these spatial patterns in ecosystems are not static, but change in the course of time.

There is a wide variety of mathematical models which deal with ecological questions. The purpose of ecological models can be very different. Quantitative predictions are normally not possible because many factors act simultaneously on ecosystems. The central aim of Theoretical Ecology (Yodzis 1989, Wissel 1989) is to get an understanding of the essential functional relations and mechanisms and to find general trends with the help of models. This is achieved by concentrating on a specific question. Only key factors for this question are taken into account. In this way the model remains plain and can provide an understanding. By this idealization and abstraction it is possible to find results which are general to a certain extent.

In this paper we use cellular automata for describing the dynamics of spatial patterns in ecosystems. That means we subdivide space into a grid of squares and time into discrete intervals. This fits to the way of collecting ecological data. In this approach rules can be used instead of mathematical equations. This makes the communication with field ecologists easier and provides a simple possibility for including empirical knowledge into the model. In this paper we show two examples which demonstrate that this simple approach is suitable to explain complex temporal and spatial behaviour of ecosystems.

2. Spreading of a Population

The first example is based on the following field data. The tephritid fly Urophora cardui is on the way to colonize Northern Bavaria, a previously unoccupied area (Schlumprecht 1989, Zwölfer 1982). This species reproduces only on the creeping thistle Cirsium arvense

Springer Series in Synergetics, Vol. 62 **Interdisciplinary Approaches to Nonlinear Complex Systems** – Eds.: H. Haken and A. Mikhailov

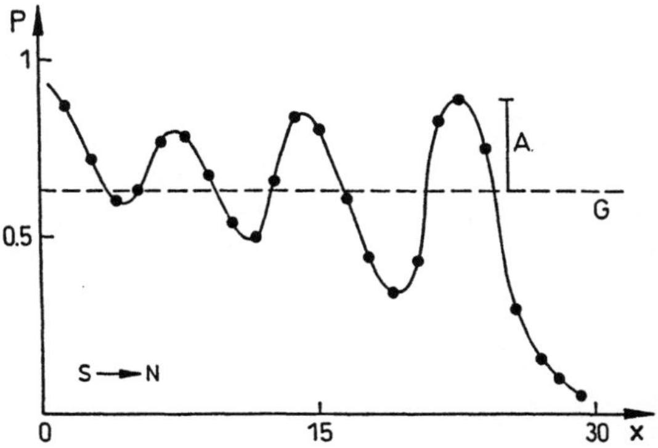

Fig. 1. Percentage P of thistle sites with populations of *Urophora cardui* versus location x of the stripes. Dispersal structure from south to north with maximum amplitude A and equilibrium G (for the year 1987). The high value of P for small x is caused by unusual local circumstances and in so far unimportant for our investigations.

(Zwölfer 1979, Lalonde and Shorthouse 1982). As a measure of density the number of thistle sites occupied by this species per square of the grid was determined. This number (density) does not decrease monotonically toward the spreading front as is often found for ecological invasion processes. Instead a pattern as shown in Fig.1 is found which moves from south to north. Details can be found else where (Jeltsch et al. 1992).

A simplified description of the population dynamics is modelled in the following way. As we want to model the spreading in one direction only the discretization of the space in one direction is sufficient. Let $P_j(x)$ be the proportion of thistle sites in the stripe x in the year j, which are at stage 1. This stage appears in the year after oviposition at an unoccupied site. $P_{j-1}(x)$ is the proportion of sites in stripe x which was in stage 1 in the year $j-1$ and which is in stage 2 in the year j. Analogously $P_{j-2}(x)$ is the proportion of sites which is in stage 3 in the year j. Here we use the empirical observation that on the average this stage 3 describes an unoccupied site. That means that a local population of this site becomes extinct two years after the colonization. Although the biological reason is not completely clear this information is sufficient for our modelling. $Q_j(x)$ is defined as the proportion of sites in stripe x that is unoccupied in the year j.

Thus our model equations are:

$$Q_j(x) = P_{j-2}(x) + \left(1 - \varphi_{j-1}(x)\right) Q_{j-1}(x) \tag{1}$$

$$P_j(x) = \varphi_{j-1}(x)\, Q_{j-1}(x) \tag{2}$$

with $\varphi_j(x)$ being the probability of the colonization of an unoccupied site in stripe x at year j. Let $f(k)dt$ be the probability that an occupied site colonizes over the distance k during the time interval dt, multiplied by the number of thistle sites per stripe. The number of occupied sites in the year j is

$$P'_j(x) = P_j(x) + P_{j-1}(x) \tag{3}$$

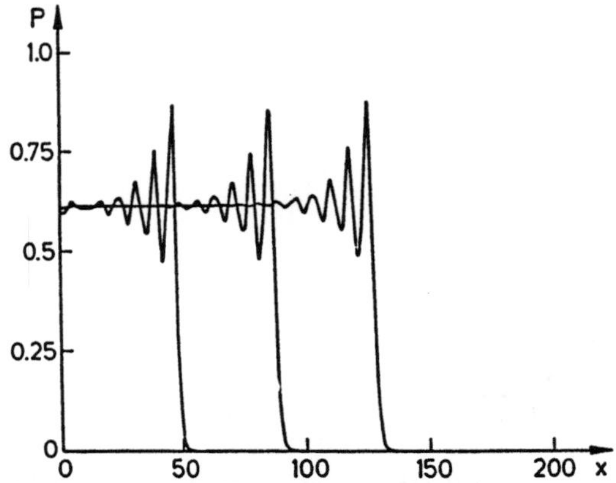

Fig. 2. Percentage P of thistle sites with existence of *U. cardui* galls in stripe x (dispersal structure). Numerical solutions of the model equations (1) and (2). $a = 3$; $b = 0.3$;$\mu = 0.4$ The three curves are the structure at three different times with a distance of 15 years. That means that the dispersal structure is propagating from left to right with constant shape and velocity.

Thus, the probability $\varphi_j(x)$ that an unoccupied site in stripe x is colonized by any occupied site is

$$\tilde{\varphi}_j(x)\,dt = \sum_k f(k)\,P'(x+k)\,dt \tag{4}$$

The probability that an unoccupied site in stripe x is not colonized during the time t is

$$W(t) = \exp\big(-\tilde{\varphi}_j(x)\,t\big) \tag{5}$$

let T be the period of time available for colonization during one year. Then the probability $\varphi_j(x)$ of the colonization of an unoccupied site in stripe x at year i is

$$\varphi_j(x) = 1 - \exp\big(-\tilde{\varphi}_j(x)\,T\big) \tag{6}$$

As random walk of moving females can be assumed we take

$$f(k) = a\exp(-bk^2) \tag{7}$$

Here a is a measure for the frequency of dispersal.

The model (1) to (7) can easily be solved by numerical iteration. We start with a distribution where occupied sites are on the left side of the x-axis and the sites on the other side are empty. For a wide range of the parameters a and b we obtain a spatial pattern as in Fig.2 which moves with constant velocity and shape from left to right. We find that the number of stages of the populations (so far 3) is almost identical with the time of periodicity of the damped oscillation in Fig.2. The general shape of the results (Fig.2) agrees very well with the field data (Fig. 1).

We define the damping as the logarithm of the quotient of two successive amplitudes. This and other characteristics are used for a rough quantitative comparison as shown

Table 1. Comparison of the characteristics of the dispersal structure: model results and field data of *Urophora cardui*
basic model: a = 1.4; b = 0.3
modified model: a = 3; b = 0.3; μ = 0.4

	field data	basic model	modified model
velocity	3.9 km/y	3.9 km/y	3.9 km/y
wavelength	10...12 km	10...12 km	10...12 km
equilibrium	60%	65%	61%
1. amplitude	25%	23%	25%
damping	0.6...0.9	0.14	0.85
periodicity	2.6...3.1 y	3 y	2.9 y

by the first two columns in Table 1. Values of the velocity and the wave length are used for fitting the parameters a and b of our model. A surprisingly good agreement is found for the other independent characteristics. Only the damping of the model result is too small.

Therefore we look for a modification of our model which increases this value. The crucial point turns out to be the fact that local populations are not always extinct exactly 3 years after colonization. Extinction can appear earlier. This is taken into account be including a probability μ that a population is extinct during one year. This modified model (Jeltsch et al. 1992) provides the same qualitative results. But the quantitative relations are change. In the third column of Table 1 the corresponding characteristics are shown. There the time of periodicity, the maximum amplitude and the wave length are used for fitting the parameters a, b and μ. Now, with this modification of the model the agreement of the other characteristics is very good.

It has to be emphasized that for the most ecological models only qualitative agreement with field data can be achieved. As here a very good quantitative description is obtained we can use our result for the determination of a quantity which is difficult to measure in the field. It is the dispersal distance of a female which is related to the fitted parameter b. The mean dispersal distance is 3 to 4 km whereas a good chance for a movement over 7 to 8 km exists.

Our approach differs from the usual method for the description of spatial and temporal processes with the help of reaction–diffusion equations. A model (Murray et al. 1986) of this type was used for describing similar patterns as in Fig. 1 for the spreading of rabies. It contains 7 parameters. By a small modification and reinterpretations of the variables our model can be used for the description of rabies. But it needs only 2 parameters and is much simpler to handle and to understand. This was possible because empirical facts can easily be adopted in a cellular automata model. We believe that this method is very suitable for modelling pattern formation in ecology. This is demonstrated also by the second example.

3. Natural Forest Dieback

3.1 The Cohort Senescence Theory

There are several examples in different parts of the world where natural forest dieback appears (Mueller-Dombois 1980, Ogden 1988, Hennon et al. 1990). That means that whole stands of trees die simultaneously without air pollution or other anthropogenic influences. This phenomenon is also called stand level dieback and is especially well investigated for the Ohia forest on Hawaii (Petteys et al. 1975, Mueller-Dombois 1988). Typical patterns with areas of dead trees are found. There exists a hypothesis, the so-called cohort senescence theory (Mueller-Dombois 1988) which is based on extensive field work and which suggests that mainly three factors are responsible for stand level dieback:

1. *Predisposing factors.* These determine the environmental preconditions for finding stand level dieback.
2. *Trigger factors.* These are randomly appearing environmental stresses like flood, drought, storm, pests, disease etc.
3. *Accelerating factors.* These cause an acceleration of dieback of senescent trees by pests, fungi, diseases etc.

We represent a model which includes these factors and is used for checking if this verbal theory can provide a sufficient explanation for the empirical findings.

Our model uses the cellular-automata approach with a cell size which provides space for a single adult tree. We consider forests which are dominated by one tree species and in which seedlings do not have a chance of growing in the shadow under the canopy of the adult trees. But if an adult tree dies the omnipresent seedlings start to grow at this site. In the course of time the smaller ones die by competition. This process is called self-thinning and is not described explicitly in our model because we do not consider numbers of trees. Instead we describe the ecological state of a cell by saying that there are trees of a certain age.

We use discrete time t with a time step of 10 years. We define t_p to be the physiological age which is an inverse measure of the vitality of a tree. For young trees ($t_p < T_1$) the physiological age t_p concurs with the time t. Thus increasing time t by one time step the physiological age increases by the same amount

$$t \to t+1 \qquad t_p \to t_p + 1 \qquad \text{for} \quad t_p < T_1 \tag{1}$$

For older trees ($t_p > T_1$) the randomly appearing trigger factors considered in the cohort senescence theory may cause an increase of the physiological age by more than one step.

$$t \to t+1 \qquad t_p \to t_p + k \qquad \text{for} \quad t_p > T_1 \tag{2}$$

The effect should be the stronger the higher the physiological age i.e. the lower the vitality of a tree.

$$k = k_m(t_p - T_1)/(T_m - T_1) \tag{3}$$

where T_m is the maximal physiological age of a tree. As these trigger factors appear randomly the maximal increase k_m of physiological age is chosen from a Poisson dis-

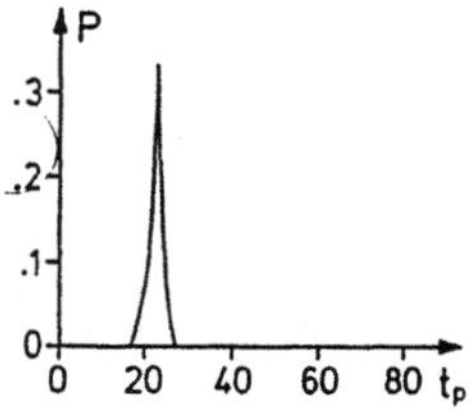

Fig. 3. Physiological age distribution after 1000 time steps. In the initial state all tree sites were synchronously at age 1. The strength of the triggering factors is $K = 4$.

tribution with the mean K. This quantity describes the mean strength of the trigger factors.

In the senescent phase ($t_p > T_2$) trees may die. Therefore we introduce a death rate μ (probability of dying per time step) which increases with t_p because of the accelerating factors (see above)

$$\mu = (t_p - T_2)^q/(T_m - T_2)^q \tag{4}$$

The definition is made in way which gives $\mu = 1$ for the maximal physiological age T_m. We use $q = 2$ which gives rise to a small increase of the death rate at the beginning of the senescent phase and a strong increase at the end. After the death of a tree the life cycle starts again with seedlings of the physiological age $t_p = 0$.

The predisposing factors of the cohort senescence theory determine the environmental conditions. Here it is essential that this life cycle of one tree species is not disturbed by other tree species. On Hawaii Ohia trees colonize new lava fields giving rise to a cohort of trees of the same age. Therefore these trees will die more or less simultaneously, i.e. a stand level dieback appears. Then the life cycle starts anew. Without any stochastic influence this cohort of tree sites would run through the life cycle synchronously for ever. But there is a variability of the time of dieback because death appears only with a certain probability, i.e. the death rate μ. Therefore one may expect that the synchrony of the life cycles of different tree sites decrease in the course of time and that after all they are completely desynchronous. But this is not the case as one can see by iterating the model equations (1) to (4) for a set of tree sites starting all with a completely synchronous age distribution. In Fig. 3 the distribution of the age classes of the different tree sites which results after 1000 time steps (10000 years) is shown. A good synchrony indicated by the moderate width of the distribution is found.

We introduce

$$S = (\ln(\text{variance}))^{-1} \tag{5}$$

as a measure of synchrony. This decreases very slowly in the course of time, but does not tend to zero, i.e to a complete desynchrony. Therefore, there must be a synchronizing factor counteracting the desynchronizing action of the stochastic death process. We investigate this question by defining the times t_1, t_2 and t_3 by

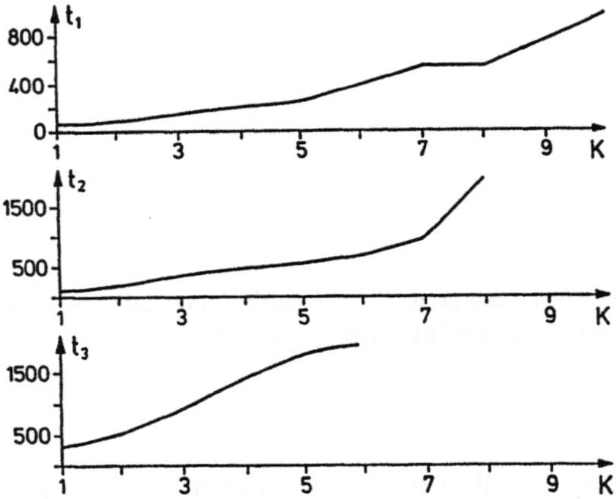

Fig. 4. Specific times, t_1, t_2, and t_3 (in time steps), when S first falls below the values 1, 0.5, and 0.3, versus the strenght of the triggering factors K.

$$\begin{aligned} S(t_1) &= 1 \\ S(t_2) &= 0.5 \\ S(t_3) &= 0.3 \end{aligned} \tag{6}$$

The longer these times t_i the slower the desynchronization and the stronger must be the synchronizing factor. In Fig. 4 these times t_i versus the strength K of the trigger factors (random environmental stress) are shown. The increase of t_i with increasing K shows that the trigger factor acts synchronizing. This becomes understandable from the following fact. By the appearance of the trigger factor the age distribution (see Fig.3) is pushed (see Equ.2) into a range of physiological age t_p where a high probability μ of dying exists. Although the trees are of different physiological age t_p they die simultaneously and at the corresponding tree sites the life cycle starts synchronously. Thus recurrent environmental stress acts synchronizing.

3.2 Spatial Effects

So far we have modelled the factors of the cohort senescence theory and have checked that they contribute to a synchronization of the age of different tree sites and consequently promote stand level dieback. But these factors cannot explain the patchy spatial patterns which are seen in nature. There must be some interactions between different tree sites. Indeed a dying tree can have a deleterious effect on the neighbouring trees. There are several possibilities like infection with pathogens, transmission of fungi or pest insects, changing of the microclimate, exposing to storms etc.

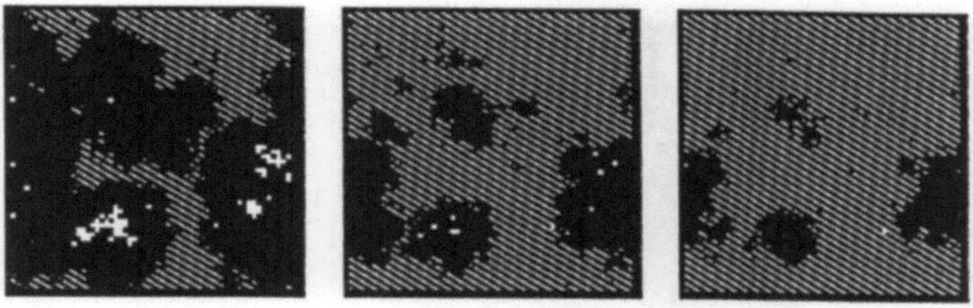

Fig. 5. Sequence of a typical patchy dieback and regeneration pattern resulting from neighbour-induced mortality and globally acting triggering factors, after 1000 time steps. Time difference between following pictures: 1 timestep. ($K = 3$; maximal distance of neighbour interaction $d_{max} = 4$; $T_1 = 25$, $T_2 = 60$, $T_m = 90$ in timesteps; 1 timesteps = 10 years). White areas stage 1 (healthy); *striped areas* stage 2 (sensitive); *black areas* stage 3 (senescent).

3.2.1 Induced Death of Neighbouring Trees

First let us consider the case in which the effect of a dying tree on the neighbours is so strong that they also die with a certain probability μ_A. This additional and independent death rate μ_A is assumed to act on old trees ($t_p > T_2$) only and this the more the older the trees are. In addition the action should decrease with the distance d of the affected tree to the affecting tree. Therefore we take

$$\mu_A = \mu_o d^{-2}(t_p - T_2) \tag{7}$$

$$\text{with} \quad \mu_o = \frac{1}{(T_m - T_2)} \tag{8}$$

and assume that there is a maximal distance d_{max} for which this neighbouring influence can act.

Including this factor (7) in our model (1) to (4) and using the method of cellular automata we find patchy spatial patterns as shown in Fig.5 where parameter values are used which describe Ohia forest on Hawaii. Nuclei of dieback appear first increasing to large patches. Indeed this type of dieback is found on Hawaii. Although sufficient field data for a quantitative comparison are not available several phenomena are found in nature which can be explained by our model.

So called *a'a'* lava flows show a worse synchronization of the age of trees, a disturbance of the patchy spatial pattern and a shorter life cycle of the trees. These lava flows have a clinker structure with small lava rocks which may not provide enough stability for the roots of growing trees. Therefore we introduce an additional death rate for young trees. This additional stochastic factor modifies the results of our model in a way which corresponds to the findings in nature: worse synchronization, disturbed spatial structure and shorter life cycle.

On very old soils the synchronization of the age of trees is also very bad. This may be related to the fact that on these riach soils other tree species (especially tree ferns) can

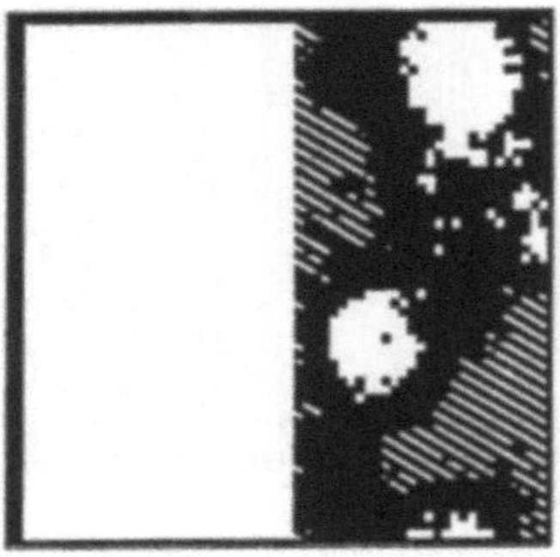

Fig. 6. Different cycles in the left and right half of the modelled forest system with homogenous initial conditions after 200 time steps. T_i(left side) $= 1.2\ T_i$(right side) $i = 1, 2m$; $K = 3$; $d_{max} = 3$

compete successfully with the Ohia trees. We take this fact into account by including a probability that no colonization by Ohia trees takes place on a site of a dead Ohia tree. Indeed this stochastic factor causes worse synchronization of the life cycles of the Ohia trees.

Field investigations have shown that the boundaries of the dieback patches sometimes coincide with the boundaries of lava fields. But sometimes a boundary of a lava field lies inside of a dieback patch. We model two adjacent lava fields by assuming different conditions (predisposing factor of the cohort senescence theory) for the left and the right half of the modelled system. Because of the different soils the duration of the phases (T_1, T_2, T_m) may be different. Our model shows that small differences of the durations do not have any influence on the results. In this case the boundary of the two parts cannot be seen in the resulting spatial pattern. But if the times T_1, T_2 and/or T_m are too different some patch boundaries of our model coincide with the separating line of the two parts (see Fig. 6).

Another explanation of these findings in nature may simply be the different colonization histories of different lava flows. This would result in an initial phase shift of the life cycles of the Ohia trees on these lava flows. A small phase shift is quickly compensated in our model. But if it is too large some boundaries of the dieback patches coincide with the separating line. These results of our model provide a possibility to check the hypothesis which suggests that the patchy pattern may be a consequence of the colonization history. But this results only in very small phase shifts of different parts of the colonized lava field and our model shows that these colonization patterns disappear quickly.

So far our model was adjusted to the Ohia forest on Hawaii. But it was constructed in a relatively idealized form without very specific details. Therefore we can try to apply it to other cases of stand level dieback. Such a case is observed for the Alaska cedar in Alaska (Hennon 1986, Hennon et al. 1990). We use estimates for the model parameters which fit to this case. There are data which provide the mean velocity of the boundaries of the expanding dieback patches (2m/year). This velocity can be obtained in our model only if the maximal distance d_{max} of the deleterious neighbouring influence is more than

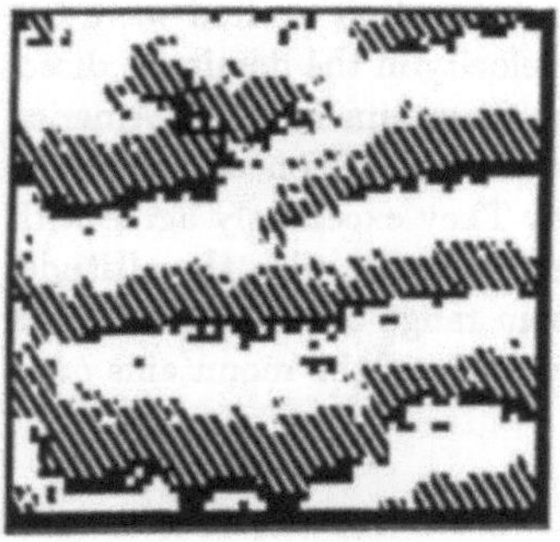

Fig. 7. Typical wave-like dieback and regeneration pattern resulting from neighbour induced triggering factors with preferende to $N - S$ direction after 2000 timesteps ($K_{\text{North}} = 5$; $K_{\text{East/West}} = 3$; $K_{\text{South}} = 1$; $q = 2$).

3 tree sites. In this case our model gives very fuzzy boundaries of the dieback patches. But this exactly is a specific property which has been found for this dieback form.

Another example of stand level dieback is found on New Zealand for the mountain beech (Ogden 1988, Jane 1986). Using estimates of the parameter values for this case we find patch sizes of the stand level dieback which are far too small compared to the findings in nature. No reasonable parameter combination produces patches which are of the enormous size as observed for the mountain beech in New Zealand. Therefore we have to conclude that a decisive factor is missing in our model. This is possibly the appearance of recurrent cyclones. They cause a very strong stress for the trees destroying parts of the canopy. We include this factor as an additional trigger factor with a strength which is at least one order of magnitude greater than the other trigger factors K. We assume that the cyclones appear with a recurrent time which varies to some degree. Our model gives very large dieback patches if this recurrent time is in the range from 120 to 190 years. This agrees with the age of the senescent phase which is 120 to 220 years. This fits very well to the estimate (Ogden et al. 1991) that the cyclones appear about every 120 years.

3.2.2 Induced Loss of Vitality

There should be cases in which the deleterious effect of a dying tree on its neighbours is not so strong that they die. A dying tree may cause a loss of vitality of its neighbours by similar factors as described at the beginning of Sect. 3.2. That means that the neighbouring tree is exposed to additional trigger factors and its physiological age t_p is increased as in (2), but with an additional mean strength K. If one tree is affected by several neighbouring dying trees the corresponding K-values may be added or the largest may be selected. But this makes no difference for the results. In our model we consider the case in which the deleterious effect of a dying tree on its neighbour depends on the direction. We choose different K-values (mean strength of the trigger factor) for the four cardinal points. The K-value for other directions are determined by interpolation.

For a certain range of the K-values with one dominant direction we find wave like patterns as shown in Fig. 7. It propagates with constant velocity in the dominant direction. This type of wave like dieback is observed on certain mountains in North America and Japan. Fitting our model parameters to the corresponding situation we can determine the velocities of the wave propagation in our model. They excellently agree with the observed values. We believe that the model parameters change with the altitude. Remember that these wave patterns are found for a certain range of parameter values only. Thus we expect to find these patterns in a certain altitude of the mountains only. Just the same is found in reality.

4. Discussion

In this paper we have shown that very simple cellular-automata models are suitable for describing spatio-temporal patterns in ecosystems. As rules instead of equations are used it is rather easy to incorporate empirical knowledge into the model. One has to start with a specific question. First one has to characterize the possible ecological states of the cells in a way which is sufficient for that question. Then one has to describe the essential key factors for this question by suitable rules. In this way one can avoid to include unnecessary details into the model.

We have constructed a general model of stand level dieback which can be adapted to specific cases. Only in a few cases quantitative field data are available. In these cases good quantitative agreement with the model is found. But also for several qualitative findings in different parts of the world our model provides good understanding and shows considerable explanatory power.

Spatial patterns and their dynamics are of great importance for ecosystems. A good example are openings in forests. They are created by the death of trees and disappear by recolonization after a while. There are many plant and animal species which are adapted to openings. These species have a chance of survival only if a sufficient number of openings are present in the forest. But there are also many other spatial structures in ecosystems which are essential for the surviving of species. As these structures are exposed to a dynamics, an ecosystem must possess a sufficient size in order to guarantee that this dynamics can run and provides the survival conditions. This point is often overlooked in nature conservation.

A model (Wissel 1991) which is very similar to the present one was constructed to describe Mid European beech forest. Rules were used which are based on empirical knowledge. As no virgin forest of this type exists in Europe this model describes what a natural forest in Mid Europe should look like. A patchy pattern with a cyclic succession is found. There is a verbal theory (Remmert 1991), the Mosaic-Cycle concept, which is based on a multitude of observations. It states that spatial patterns with patchy structures and cyclic local dynamics should appear in many ecosystems. We believe that models which are modifications of the present one should be suitable for describing these spatio-temperoral patterns of ecosystems.

5. Acknowledgement

This research was supported by the Deutsche Forschungsgemeinschaft (DFG).

References

Hennon, A.S. (1986): "Pathological and ecological aspects of decline and mortality of chamaecyparis nootkatensis in southeast Alaska", Ph. D. thesis, Department of Botany and Plant Pathology. Oregon State University, Corvallis

Hennon, A.S.; Hansen, E.M.; Shaw, C.G. III (1990): "Dynamics of decline and mortality in chamaecyparis nootkatensis in southeast Alaska", Can. J. Bot. **68**(3), pp. 651–662

Jane, G. T. (1986): "Wind damage as an ecological process in mountain beech forests of Canterbury, New Zealand", NZ J. Ecol., pp. 25–39

Jeltsch, F.; Eber, S.; Wissel, C.; Brandl, R. (1991): "Oscillating dispersal patterns of tephritid fly populations", Ecol. Model **60**, pp. 63-75

Jeltsch, F.; Wissel, C. (1992): "Modelling factors which may cause stand-level dieback in forest", in Huettl, R. and Mueller–Dombois, D. (eds.): "Forest decline in the Atlantic and Pacific region" Springer (1992)

Lalonde, R.G. and Shorthouse, I.D. (1982): "Exit strategy of *Urophora cardui* (Diptera: Tephritidae) from its gall on Canada thistle", Can. Entomol., **114**, pp. 873–878

Mueller-Dombois, D. (1980): "The Ohia dieback phenomenon in the Hawaiian rain forest" in Cairns, J.jr. (ed):"The recovery process in damaged ecosystems." Ann. Arbor Science Publ. Ann. Arbor, Mich.: pp. 153–161

Mueller-Dombois, D. (1988): "Towards a unifying theory for stand-level dieback." Geo Journal **17** 2: pp. 249-251

Murray, J. D.; Stanley, E.A.; Brown, D.L. (1986):"On the spatial spread of rabies among foxes", Proc. R. Soc. London B **229** pp. 111-150

Ogden, J. (1988): "Forest dynamics and stand-level dieback in New Zealands Nothofagus forests. Geo Journal **17** 2: pp. 225–230

Ogden, J., Fordham, R.A., Pilkington, S. (1991): "Forest gap formation and closure along an altitudinal gradient in Tongariro National Park", New Zealand, J. Veg. Sci. **2** pp. 165-172

Petteys, E.Q.P.; Burgan, R.E.; Nelson R.E. (1975): "Ohia forest decline: its spread and severity in Hawaii", USDA For. Serv. Res. Paper PSW–105.

Remmert, H. (1991) (ed): "The mosaic-cycle concept of ecosystems" Ecological studies **85** Springer Verlag, Berlin Heidelberg New York

Schlumprecht, H. (1989):"Dispersal of the thistle gallfly *Urophora cardui* and its endoparasitoid *Eurytoma serratulae* (Hymenoptera: Eurytomidae)", Ecol. Entomol., **14**, pp. 341–348

Wissel, C. (1989): "Theoretische Ökologie: Eine Einführung" Springer Verlag, Berlin Heidelberg New York

Wissel, C. (1991): "A model for the mosaic-cycle concept" in Remmert, H. (ed.) "The mosaic-cycle concept of ecosystems" Ecological studies **85** pp. 22–46. Springer Verlag, Berlin Heidelberg New York

Yodzis, P. (1989): "Introdutction to Theoretical Ecology" Harper and Row, New York

Zwölfer, H. (1979): "Strategies and counterstrategies in insect population systems competing for space and food in flower heads and plant galls", Fortschr. Zool., **25**, pp. 331– 352

Zwölfer, H. (1982): "Das Verbreitungsareal der Bohrfliege *Urophora cardui* L. (Diptera: Tephritidae) als Hinweis auf die urspriglichen Habitate der Ackerdistel (*Cirsium arvense)* (L.)" Scop. Verh. Dtsch. Zool. Ges. 298

Characterization of Temporal and Spatio-temporal Chaos

A. Babloyantz

Service de Chimie-Physique, Université Libre de Bruxelles,
CP 231 - Campus Plaine, Boulevard du Triomphe, B-1050 Bruxelles, Belgium

Introduction

Since the advent of the concept of deterministic chaos, several methods have been proposed for their characterization from experimental time series. The evaluation of various dimensions and the spectrum of Lyapounov exponents are among the most popular techniques. These methods, in general, have been developed and tested on ideal cases where the dynamics and the attractors are well defined and noisless. In these cases reliable parameters of system's dynamics could be obtained from time series analyses. Thus in principle we have tools for probing temporal chaos. Such a possibility has triggered a wealth of publications in such diverse fields as chemistry, physics, hydrodynamics, climatology, physiology and economics.

On the other hand spatio-temporal chaos, as encountered in many interesting systems, and to our best knowledge few reliable techniques has been proposed for their quantitative evaluations.

In this paper we discuss the much controversial problem of the existence of chaotic attractors in brain dynamics. We also propose the non linear time serie analyses as a tool for probing spatio temporal chaos. Such an endeavor brings us to the study of neural networks and the effect of number of elements, and their connectivity on system's dynamics.

Relevance of deterministic chaos in Physiology

The techniques of nonlinear time series analysis were applied for the first time in physiology for the study of dynamics of electroencephalograms (EEG) by Babloyantz, Nicolis and Salazar (1985) and the dynamics of single neurons by Rapp, Zimmerman, Albano and de Guzman (1985).

These works generated a great enthusiasm and triggered the hope that at last we are in possession of a technique that might give a handle for probing untreatable problems such as for example the nature of cerebral activity.

Since that time nonlinear time series analysis has been applied with confirmed success only to few selected problems. On the other hand in view of deceptively simple algorithms it has also been used in very diverse contexts by various researchers not

Springer Series in Synergetics, Vol. 62 **Interdisciplinary Approaches to Nonlinear Complex Systems** – Eds.: H. Haken and A. Mikhailov

very familiar with the basic ideas underlying the algorithms. Thus in some instances the techniques have not been used in the limits of their applicability; the important dynamical parameters have been estimated carelessly ; and erroneous conclusions have been drawn from insufficient data or for lack of theoretical background.

There have been enthusiastic claims that the various algorithms could be incorporated into push button softwares. Others attribute to them diagnostic value which is claimed can be used in cardiology clinics as well as in monitoring psychiatric disorders or cerebral activity of combat pilots in action.

In this short paper, there is no room for a deep discussion of these studies therefore I take the liberty of not mentioning the numerous relevant references. I will stick to the work of our group in order to illustrate the difficulties and the pitfalls, but also the indisputable success of the nonlinear time series analysis.

As should be the case in any healthy scientific debate there have also been many criticisms. The proponents of this negative approach can be divided into two broad categories. The first group of critics have a second- hand opinion about the subject. Their opinion is not based on solid grounds for lack of personal experience in the matter.

In the second category we find researchers with a solid background and understanding of deterministic chaotic dynamics. They know all the pitfalls one can encounter by applying the dynamical time series analysis to non stationary, noisy and uncontrolled data as is the case for EEG and electrocardiograms (EEG). An excellent and balanced view of the methods and their shortcomings can be found in a review paper by Grassberger, Schreiber and Schaffrath (1991). As these authors point out wisely, in order to discriminate between determinism and randomness - the central issue in much of this analysis - one should not only limit the scope to scaling laws holding in the limit of infinitely fine resolution. I will take such an approach and by common sense arguments show that several stages of human EEG follow a "non random" dynamics. For lack of a better mathematical vocabulary and concept, and in view of their resemblance to deterministic chaotic dynamics I shall conclude that one is entitled to think of these states as chaotic.

The exact relationship between the neuronal activity and the EEG is not known. However it is well established that the latter reflects in some way an average activity of neural masses. The morphology of several stages of human EEG is so characteristic that it has been used over a century as a diagnostic tool, as their general form remains unchanged for all individuals in a given state.

In an individual with eyes open (β waves) electrical activity does not have any apparent regularity : the low amplitude high frequency waves are reminiscent of random noise. However as soon as eyes are closed, the activity switches to a more regular, lower frequency and higher amplitude waves, known as "α rhythm". The phase portrait is a round shape structure with a hole in the middle. Just from visual inspection, it is thus obvious that the time series does not represent random noise. Under ideal conditions and extreme care, it is possible to obtain sufficiently clean and long data sets from the so called "good α producers".

Babloyantz and Destexhe (1987,1988) estimated the correlation dimension and Gallez and Babloyantz (1991) the Lyapounov exponents of these data sets. The same analysis was performed for β waves. The difference between the two states, which arises by merely closing the eyes, was enormous. Where β waves looked very high dimensional the α waves showed a $D_2 = 6.23$. In the sequel I would like to call these quantities "dimension", keeping however in mind that this definition may not have the same meaning as in mathematics.

There is no doubt that something has happened here. The coherence has increased dramatically and simultaneously, the "dimension" dropped also dramatically. As deep sleep sets in, the morphology of the waves changes again. The higher amplitude and lower frequency waves, as compared to α waves, are less regular, but still exhibit a well defined morphology. The dimension drops to values above four.

In the case of severe pathologies, such as the Creutzfeld-Jacob coma and "petit mal" epilepsy, the change is even more dramatic. In the case of Creutzfeld-Jacob coma in the terminal stages of the diseases, many hours of steady and practically noise free recordings are available. The waves seem as highly reproducible pseudo periodic activity.

From the inspection of the time series, two possibilities emerge. Either we are in the presence of noise contaminated periodic activity or a low dimensional deterministic chaos underlies the dynamics.

When testing for stationarity with the help of recurrence plots (Eckmann and Ruelle 1987) we could demonstrate the presence of a hidden superimposed periodic activity of 59 sec. over (Babloyantz 1991), the time series which has a pseudo periodicity of 1 sec. The phase portrait, constructed from this time series, sampled in regular time intervals, is highly inhomogeneous : the majority of points cluster in a small region of phase space, but other regions are also visited with a smaller probability. The overall portrait shows a well-structured attractor (see Fig. (1)). Let us note that this is often the case in physiological data.

The dynamical parameters were measured for this global inhomogeneous attractor as well as from the two separate parts with different density points. The last procedure was more suitable for evaluating Lyapounov exponents (Gallez and Babloyantz 1991). We thought the procedure legitimate for characterising in some way the difference between the two parts as long as the results are not interpreted as scaling laws of a unique attractor. All measured parameters point in the direction of a deterministic dynamics and not a noisy limit cycle. The later possibilities would have been unlikely on physiological grounds, since it will require in-phase activity of 10^{10} inter- connected dynamical units.

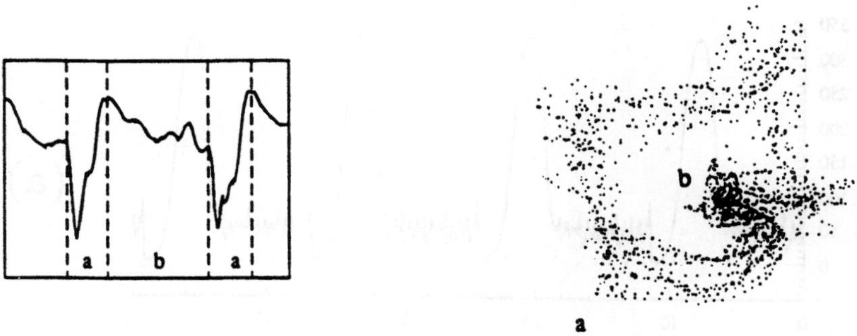

Fig. (1) : Time series and a two dimensional non homogeneous attractor from Creutzfled-Jacob coma.

When experimental time series show such "inhomogeneous" attractors, several questions arise. Are the usual techniques of non linear time series analysis applicable to inhomogeneous attractors? Are we in the presence of a single dynamics or the systems switches alternatively between two different dynamics?

In answering the first question, we believe that the straightforward use of the algorithms for the evaluation of D_2 in the case of an inhomogeneous attractor gives an average value which is different from the values one could obtain by considering each separate parts of Fig. (1) as an independent entities. We propose that a better way to deal with such attractor is to evaluate the Lyapunov exponents and correlation dimensions of each part separately. We think such a procedure is legitimate for characterizing in some way the difference between the two parts of the attractor as long as the results are not interpreted as scaling laws of partial attractors. In the case of Fig. (1), D_2 of the global attractor is $D_2 = 3.8 \pm 0.1$ whereas the so called semi-local values are $D_2 = 2.1 \pm 0.1$ for the section (a) and $D_2 = 5.1 \pm 0.2$ for the part (b) of the attractor.

An answer to the second question may be found by constructing dynamics which generate inhomogeneous attractors.

Let us consider the following set of coupled non linear differential equations :

$$\begin{aligned} \frac{dX}{dt} &= \sigma(Y - X) \\ \frac{dY}{dt} &= r\,X - Y - X\,Z \\ \frac{dZ}{dt} &= -Pr\,Z + X\,Y - c(t)\,(Z - Z_0) \end{aligned} \qquad (1)$$

One recognizes immediately the Lorenz equation which comprises an additional term $c(t)\,(Z - Z_0)$. Here $c(t)$ is a periodic function of the form $c(t) = 10/(1 + \exp[10 + 10\sin(t)])$ and is constructed in such a way as when c is practically zero

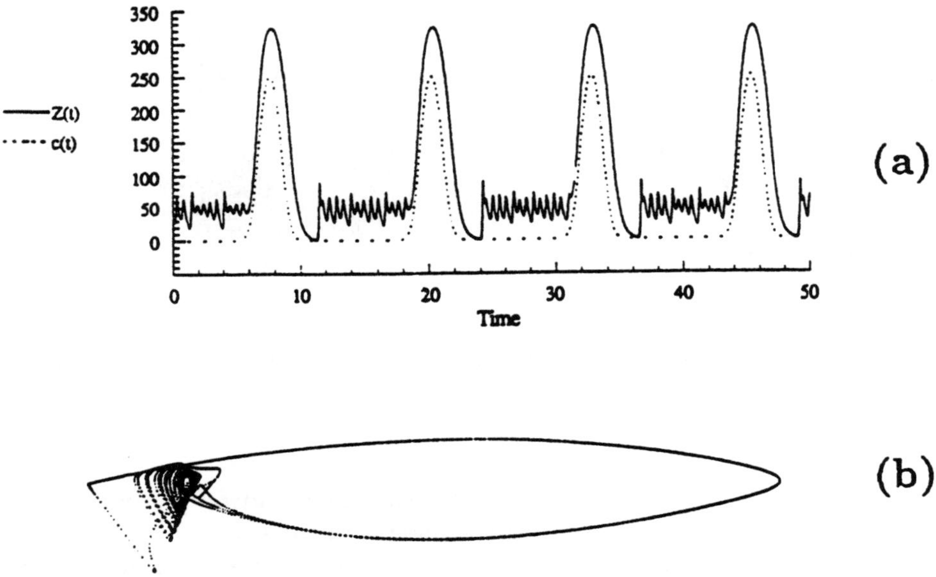

Fig. (2) : (a) Time series of Z variable from modified Lorenz attractor. (b) The corresponding non homogeneous attractor.

the system is pulled toward a chaotic attractor. However as c increases the system is drawn toward a periodic orbit.

The time evolution of variable Z is depicted in Fig. (2) where one sees that chaotic dynamics are followed regularly by a burst of periodic activity.

The phase portrait constructed from this time series shows indeed two distinct regions of chaotic and periodic behaviour.

If the correlation dimension of such an attractor is evaluated by considering the two regions separately one find a value of $D_2 = 1.01 \pm 0.01$ for the periodic part whereas the dimension is $D_2 = 2.04 \pm 0.03$ for chaotic end of the phase portrait (Destexhe, Ph.D. dissertation).

From what had been said previously and from our present knowledge of non linear dynamics we have the choice between two entirely different scenarios as regard to EEG activity.

1) The EEG of a healthy subject is characterized by Gaussian random noise. This noise is such that it can be classified in a single individual into different categories and is similar in different individuals. On the contrary in pathologies the activity switches to a noisy limit cycle. Such a possibility is very unlikely.

2) The other alternative, which is supported by the work of our group and many others is that the EEG follows a deterministic dynamics in all cases cited above. This view is supported by the great difference between coherence of certain stages of EEG and less coherent ones such as β rhythm. It is true that, even in coherent states, "deterministic chaos" appears through some highly averaged properties such as attract dimension. More detailed information concerning fine scale dynamics remains unavailable. Despite of these shortcomings in our opinion in this decade of brain, the dynamical approach to cerebral activity is the only way for probing brain dynamics. Indeed, as imperfect as the non linear time series analysis of physiological data may be, we feel, it has opened new avenues of research in a field where only the experimental or a psychological methods were available so far.

Measuring spatio-temporal chaos

Recently there has been a great deal of interest in the study of spatio- temporal chaos as they appear in many fields such as chemical media, hydrodynamics, laser physics, as well as in neural network models of artificial and biological neurons. Unfortunately so far there are no reliable methods for measuring the degree of coherence of a system exhibiting spatio-temporal chaos. We believe that non linear time series analyses could be of some help in this area. We propose the following procedure which will be illustrated on a particular network which mimicks the activity of the cerebral cortex.

Let us consider a network of N excitatory and M inhibitory interconnected neurons. The dynamics of the network is described by the following equations :

$$
\begin{aligned}
\frac{dX_i}{dt} &= -\gamma(X_i - V_L) - (X_i - E_1)\left[\sum_k \omega_{ki}^{(1)} F(X_k(t-\tau_{ki})) + T_i\, g(x)\right] \\
&\quad -(X_i - E_2)\sum_l \omega_{li}^{(2)} F(Y_l(t-\tau_{li})) \\
\frac{dY_j}{dt} &= -\gamma(Y_j - V_L) - (Y_j - E_1)\sum_k \omega_{kj}^{(3)} F(X_k(t-\tau_{kj})) \\
&\quad -(Y_j - E_2)\sum_l \omega_{lj}^{(4)} F(Y_l(t-\tau_{lj})) \\
&\qquad i,k = 1 \ldots N \quad , \quad j,l = 1 \ldots M
\end{aligned}
\tag{2}
$$

Here X_i and Y_j are respectively the postsynaptic potentials of excitatory and inhibitory neurons. γ is the inverse of the time constant of the membrane and $\omega_{ji}^{(1)}, \omega_{ji}^{(2)}, \omega_{ji}^{(3)}, \omega_{ji}^{(4)}$ are the synaptic weights between two excitatory, excitatory and inhibitory and inhibitory-inhibitory cells. V_L is the resting potential whereas E_1 and E_2 are equilibrium ionic potentials. F is a sigmoïdal function which takes account of a delay term describing propagation times in the network.

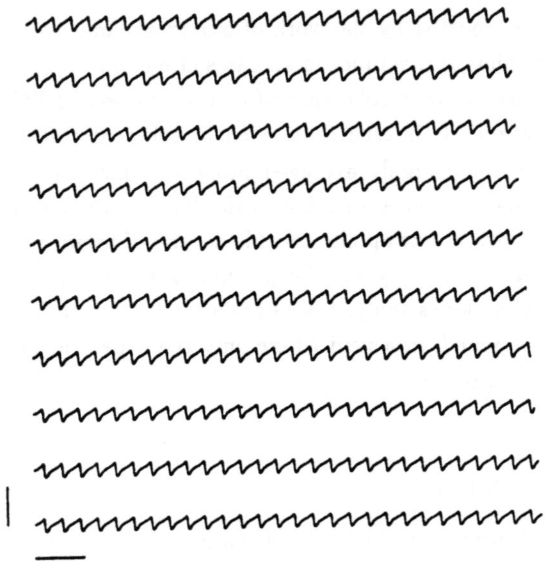

Fig. (3) : Periodic activity of 10 cells from a network of neurons. $N = 100$, $M = 25$, $\Omega_1 = 14.25, \Omega_2 = \Omega_3 = 12.5, \Omega_4 = 0$.

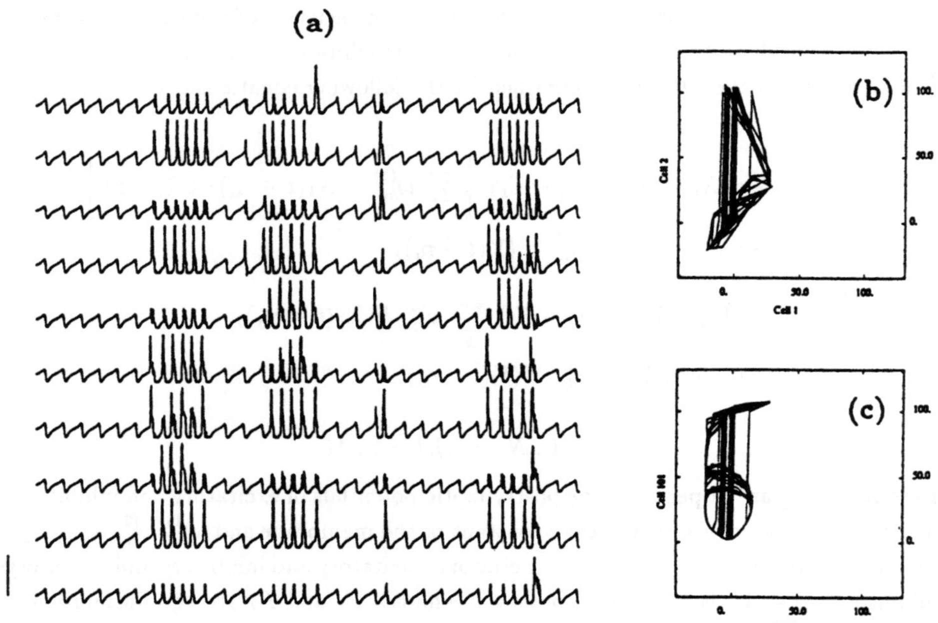

Fig. (4) : Intermittant activity of the network of Fig. (3). Here $\Omega_1 = 14.4$.

(b)

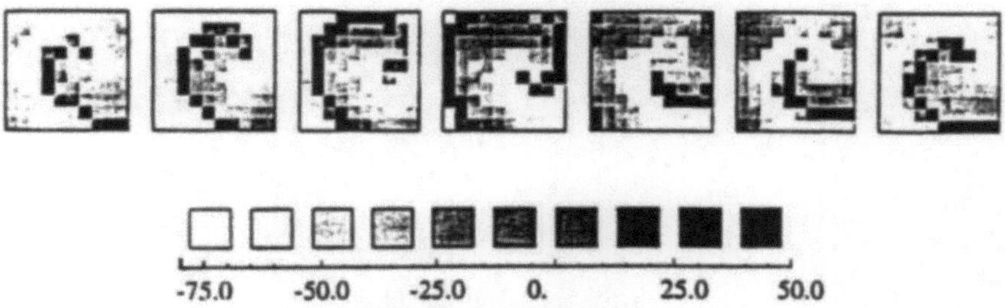

Fig. (5) : Spiral activity is seen in the network of Fig. (3) if $\Omega_1 = 15$.

If one fixes all parameters of the model and increases excitatory to excitatory connections gradually, the activity of the network is switched between different states. It is important to note that the network activity is strongly dependent on the size of the network.

Let us first consider a moderatly small network with $N = 100$ and $M = 25$. Moreover $\Omega_1 = \sum_k \omega_{ki}^{(1)} = 14.25$, $\Omega_2 = \sum_l \omega_{li}^{(2)} = \Omega_3 = \sum_k \omega_{kj}^{(3)} = 12.5$ and we neglect inhibitory to inhibitory connections. The activity of ten excitatory cells are shown in Fig.(3). One observe a periodic activity throughout the network.

As Ω_1 is increased one observes spatio-temporal intermittency in the same ten cells as seen in Fig.(4).

A further increase in the value of $\Omega_1 = 15$ gives rize to spiral type of activity (see Fig.(5)). In other range of parameter values target waves are seen in the system.

Now if we increase the size of the system, keeping all parameters constant the activity switches to a spatio-temporal chaotic state. However if we increase the connectivity of the system by considering second neighbour interactions, the system switches back to an oscillatory state.

An example of a chaotic network is given in Fig.(6) where $N = 6400$ and $M = 1600$ cells are connected, via second neighbour interactions. The network exhibits spatio-temporal chaos.

Our various simulations showed that an appropriate rate of connectivity radius over the number of cells must be reached before the onset of the spatio-temporal chaos. The spatially averaged value of the network activity furnishes a time series that can be analyzed by the usual techniques of non linear times series analyses. One sees that as the coherence of the network increases the value of D_2 diminishes thus giving a quantitative handle for characterization of spatio-temporal chaos.

To see this more clearly, let us introduce a periodic input into 2% of the excitatory cells of the network. Fig.(7) shows a radical charge in the spatial activity of the network.

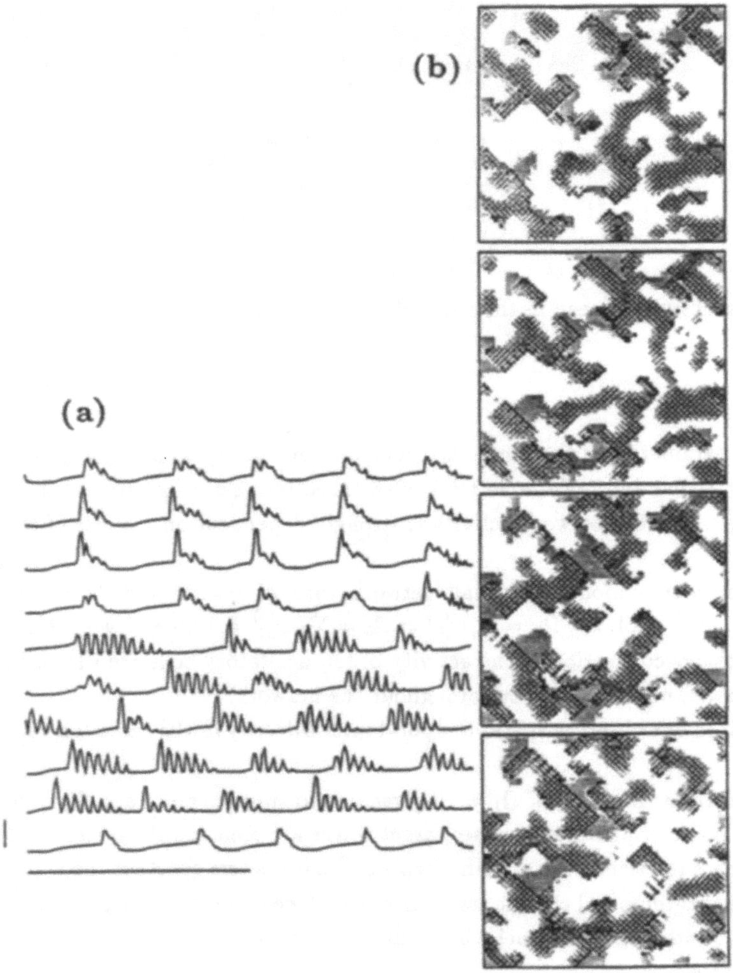

Fig. (6) : Chaotic activity in a network of $N = 6400$, $M = 1600$ cells although second neighbour interactions are considered.

A large patches of cells in synchrony appear and disappear in time showing an obvious increase in the coherence of the system.

Fig.(8a,8b) shows the spatial averaged time series obtained respectively from network of Fig.(7) together with their two dimensional phase space.

One sees immediately that an obvious attractor appears in the more coherent network. The evaluation of the correlation dimension shows a value of $D_2 = 3.6 \pm 0.1$ for

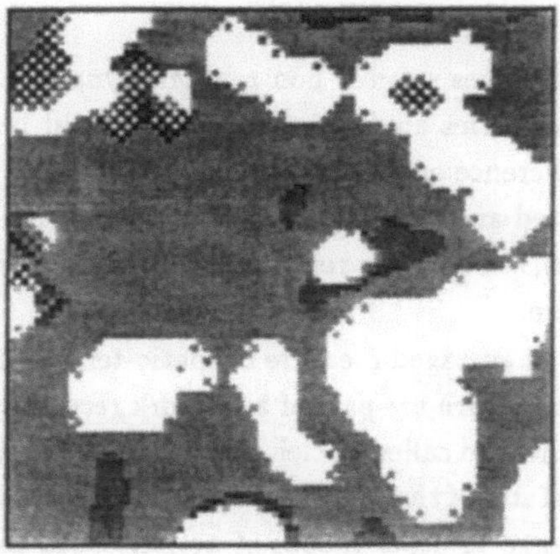

Fig. (7) : Periodically stimulated excitatory neurons (2%) introduce Coherence in the network.

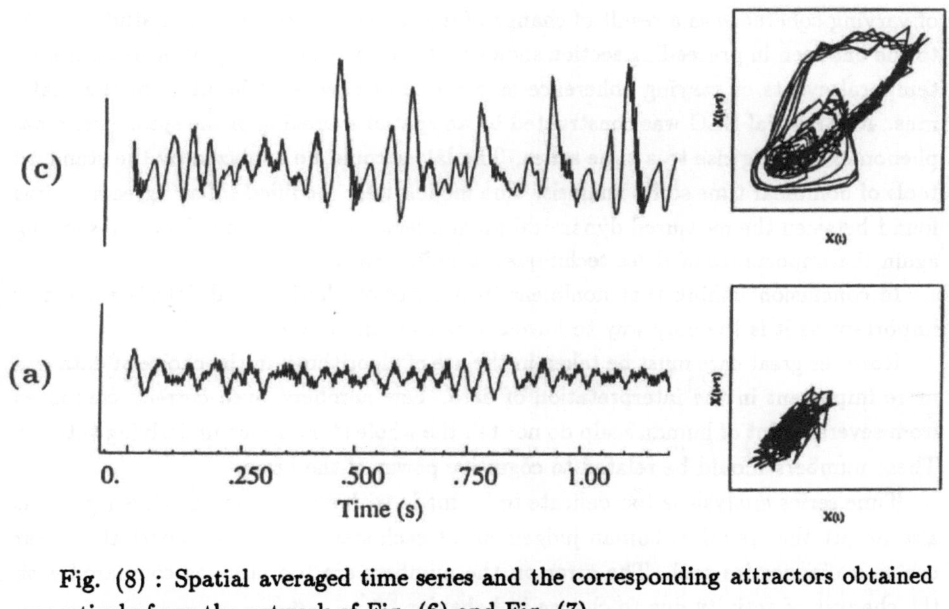

Fig. (8) : Spatial averaged time series and the corresponding attractors obtained repectively from the network of Fig. (6) and Fig. (7).

this attractor. No saturation could be found for the attractor obtained from the less coherent network. With this example we see that correlation dimension could be used as a tool for quantification of spatio-temporal chaotic activity.

Discussion

The characteric form of certain stages of brain waves suggest non random dynamics. If care is taken in applying the non linear time series analyses, parameters could be evaluated which indicates the degree of the coherence of the EEG signal.

Assuming that dimensions could be evaluated and there is indication for the presence of deterministic chaos, the question arises, what is the relationship between the dimension and the behavioral states of the brain.

The EEG expressed as a temporal event is an averaged measure of spatio-temporal activity of millions of neurones. These neurones in turn are part of a network receiving input from various parts of the brain. Among these so called nuclei, the thalamus plays an important role in controlling the behavioral states of the brain. It shows autonomous oscillatory behavior which moreover could change to other modes of activity such as fixed points, burst activity or oscillatory activities of varying frequency as a result of input into the system.

Destexhe and Babloyantz (1991) have constructed recently a simplified model of thalamocortical activity which accounts for the changes in cortical networks as a results of input to the thalamus. The cortical activity appears as spatio-temporal events of varying coherence as a result of change of input from the thalamus. A study similar to the one seen in preceeding section shows that the cortical activity appears as spatio-temporal events of varying coherence as a result of change of input from the thalamus. An artificial EEG was constructed by an spatial averaging of the spatio-temporal phenomena giving rise to a time series. The latter could be analyzed by the standard tools of nonlinear time series analysis. The models were modified till an agreement was found between the measured dynamical parameters and the computed values, showing again the importance of these techniques in brain research.

In conclusion I think that nonlinear time series of physiological data is extremely important as it is the only way to have access to brain activity.

However great care must be taken in the use of algorithms, in the choice of data and more important in the interpretation of data. Few numbers, even correct, computed from several point of human scalp do not tell the whole story about underlying activity. These numbers should be related to cognitive power of the brain.

Time series analysis is too delicate to be made with an automated procedure. It is also an art that requires human judgement at each step. I am also sceptical in their use as a diagnostics tool. The error on the numbers are too large as compared with the changes of activity due to change in behavior.

Research is non linear time series analysis must continue and there is a need for new concepts and new algorithms.

References

[1] A. Babloyantz, C. Nicolis and M. Salazar, Evidence for chaotic dynamics of brain activity during the sleep cycle. *Phys. Lett. A* **111**: 152-156 , 1985.

[2] P.E. Rapp, I.D. Zimmerman, A.M. Albano, G.C. de Guzman and N.N. Greenbaun: Dynamics of spontaneous neural activity in the simian motor cortex: the dimension of chaotic neurons, *Phys. Lett.* **A 110**, 335-338 (1985).

[3] A. Babloyantz and A. Destexhe, Low dimensional chaos in an instance of epileptic seizure. *Proc. Natl. Acad. Sc. USA* **83**: 3513-3517 , 1986.

[4] A. Babloyantz & A. Destexhe: Strange Attractors in the Human Cortex, in Temporal disorder in **human oscillatory systems**, Eds. L. Rensing, U. an der Heiden and M.C. Mackey, Springer Series in Synergetics 36,48 (1987)

[5] J.P. Eckmann, S.O. Kamphorst & D. Ruelle: Recurrence Plots of dynamical systems, *Europhys. Lett.* **4**, 973-977 (1987).

[6] A.Babloyantz & A.Destexhe: The Creutzfeld-Jakob Disease in the Hierarchy of Chaotic Attractors, in **From Chemical to Biological Organization**, Ed by M.Markus, S.Mller and G.Nicolis, Springer-Verlag (1988).

[7] A. Destexhe, J.A. Sepulchre & A. Baboyantz: A Comparative Study of the Experimental Quantification of Deterministic Chaos, *Physics Letters* **A 132**, 101 (1988)

[8] A. Babloyantz: Some Remarks of Nonlinear Analysis of Physiological Time Series, in: **Measures of Complexity and Chaos**, Eds. N.B. Abraham, A.L. Albano, A. Passamente and P.E. Rapp, NATO ARW Series, Plenum Press , 339 (1989)

[9] A. Babloyantz : Estimation of Correlation Dimensions from Single and Multichannel Recordings. A Critical View. Ed. E. Basar. Springer Series in Synergetics. Berlin (1989).

[10] A. Destexhe & A. Babloyantz, Pacemaker-induced coherence in cortical networks, *Neural Comp.* **3**: 145-154 , 1991.

[11] D. Gallez and A. Babloyantz, Predictability of human EEG: a dynamical approach, *Biol. Cybern.* **64**: 381-391 , 1991.

[12] A. Babloyantz, Evidence for slow brain waves: a dynamical approach, *Electroenc. and Clin. Neurophys.* **78**: 402-405 , 1991.

[13] D. Gallez and A. Babloyantz, Lyapunov exponents for nonuniform attractors *Phys. Lett. A* **161**: 247-254 , 1991.

[14] P. Grassberger, T. Schreiber and C. Schaffrath: Nonlinear time sequence analysis, *Int. J. of Bif. Chaos* **1**, 521-543 (1991).

Attractor-Ruled Dynamics in Neurobiology: Does it Exist? Can it be Measured?

R. Cerf

Laboratoire d'Ultrasons et de Dynamique des Fluides Complexes, Unité CNRS n°851, Université Louis Pasteur, Strasbourg, France

1 Introduction

Two concepts of Physics, perhaps the most significant ones of this half-century, self-organization and chaos, are widely held to be of key-importance in understanding aspects of Neurodynamics, and in particular brain dynamics.

When, at a higher degree of non-linearity, a complex self-organized system becomes chaotic, increased richness in behaviours exists potentially, as well as possible jumps from one behaviour to another. Even in a stationary regime, a chaotic system is unstable, being ruled by a strange attractor, which is a fractal object of non-integer dimension. The system then possesses a lability of a particular kind, and capabilities of exploring phase space. All these properties, indeed, seem attractive in neurobiological studies.

However, extreme care must be exercised when, as in our case, the purpose is to interpret electroencephalographic (EEG) signals. Complex systems, when they become self-organized, can be described by a small number of collective variables. Often, the evolution through time is studied by writing a set of first order differential equations for these variables, defining an autonomous dynamical system. Autonomous means that the equations do not contain the time explicitly.

Furthermore, we are in a way compelled to study stationary regimes. These regimes are of a particular significance in dynamical studies, and are so much easier to treat theoretically, that in most studies one tries to catch a neurobiological system when it is stationary. In other words, an autonomous dynamical system is an ideal model in these studies, which we hope, furthermore, to observe in a stationary regime for a sufficiently long time, and it will be one of my goals to explain what a sufficiently long time actually means.

In a monopolar or dipolar EEG recording the electric signal stems from a cortical territory (our system), and this certainly interacts in a complicated way with other parts of the brain. We must, therefore, expect to be faced with difficulties reflecting the non-ideality of our experimental system, and the non-stationarity of its evolution through time.

A good question, which was raised by Babloyantz et al. [1] and by Rapp et al. [2], is whether in the genesis of EEG signals, which exhibit both some regularity and a chaotic appearance, deterministic chaos has a role to play. We need not ask whether EEG experiments manifest self-organization. This follows on from the observation of oscillatory patterns.

On the other hand, we need not restrict the question to chaos. Let us, therefore, rephrase the question in both more general and more restricted terms, as follows :

"Is there evidence for attractor-ruled dynamics in Neurobiology?"

The answer is yes. Examples of repetitive responses in single nerve fibres are well known. Also, in human EEG analysis, Babloyantz and Destexhe [3] have analyzed an epileptic seizure recording that is a good candidate for low-dimensional attractor behaviour, of a dimension of the order of 2.

However, it is known that the difficulties encountered in attractor characterization increase dramatically with the attractor's dimension. Huge difficulties are encountered even for the lowest dimension of 4 that was reported for healthy humans. So let us ask :

"Is it possible from a monopolar or a dipolar EEG time series to characterize attractor behaviour for a healthy human?"

2 Detecting and Characterizing Attractor Behaviour

First a phase trajectory must be constructed, in our case from a single time series. As suggested by Ruelle [4], this can be done by using the experimental time series :

$$X(t_1), X(t_2), \ldots, X(t_N)$$

to generate the following delayed time series :

$$\begin{array}{ccc} X(t_1+\tau) & \ldots & X(t_N+\tau) \\ \vdots & & \vdots \\ X(t_1+(n-1)\tau) & \ldots & X(t_N+(n-1)\tau) \end{array}$$

where τ is the delay time (or lag time); here X is the EEG voltage. To each time t_i, $1 \leq i \leq N$, there now corresponds a point, of coordinates $X(t_i) \ldots X(t_i + (n-1)\tau)$. This point describes a phase trajectory in an n-dimensional phase space.

It is supposed that the number N of experimental values is quite large; n, the dimension of phase space, must be larger than or equal to twice the attractor's dimension plus one :

$$n \geq 2d + 1.$$

Under these conditions, theorems of Whitney and of Takens show that an embedding exists between the attractor and the constructed phase trajectory. Thus, dynamic equivalence exists between them. We must therefore measure the dimension of our phase trajectory in embedding spaces of increasing dimension. If the system is low-dimensional, values of the embedding dimension can be reached for which Whitney's theorem is fulfilled. The dimension measured saturates at the trajectory's dimension, which is also the dimension d of our attractor.

The dimension of the phase trajectory remains to be measured. In the simplest procedure, one calculates its correlation dimension, using an algorithm due to Grassberger and Procaccia [5]. This is the procedure most commonly used by neurophysiologists, and the one I shall be discussing. One counts the pairs of points on the trajectory, the distance of which is smaller than a length r. One considers each point of the trajectory in turn. The number of pairs formed with one of them, P_i, is equal to the number of points other than P_i contained in the hypersphere of radius r centered on P_i : call this

$C_i(r)$. We sum up the C_i s for all points P_i, and obtain the correlation integral :

$$C(r) = \sum_i C_i(r).$$

For a curve, of dimension 1, $C_i(r)$ is proportional to r^1. For a planar object, it is proportional to r^2, and so on. Thus, for an object of dimension ν, $C_i(r)$, and therefore also $C(r)$, is proportional to r^ν, at least in a certain range of values of r :

$$C(r) \sim r^\nu,$$

where ν is the correlation exponent. In the preceding examples ν is the Euclidian dimension, but the whole procedure is valid for a fractal object of non-integer dimension. Following Rapp and his colleagues [2], we plot the logarithmic derivative $d\log C(r)/d\log r$ in function of $\log C(r)$; we call these curves "slope curves", and the family of curves for different embedding dimensions, a Rapp-plot. In this plot, a scaling range of $C(r)$ appears as a horizontal segment of an ordinate equal to the correlation exponent ν. We do this for embedding dimensions between 2 and 40 :

$$2 \leq n \leq 40.$$

For an attractor of a low dimension, when the analyzed time series is long, the saturation of ν in function of n results in the concentration of slope curves in a reinforced horizontal range, as, for example, in Fig.1 of Ref. [6]. Such a range is the attractor's signature, and its ordinate equals the attractor's correlation dimension d.

However, a number of difficulties are encountered in attempts to analyze EEG signals. First, with very few exceptions, the expected scaling ranges are not observed, probably mainly because the length of the analyzed time series is limited by the non-stationarity of the process. Instead, other scaled structures, which I shall go on to describe, must be used.

Non-stationarity confronts us with the problem of analyzing short time series. Especially for these series, a trivial artifact, which we must get rid of, appears at small values of the radius r, when the strand of the phase trajectory that goes through the center of a hypersphere has overwhelming weight. The analyzing procedure then mainly sees an object of dimension 1, and an artifactually low correlation dimension is obtained. We correct this "single-strand artifact" using Theiler's procedure [7], whereby the $w-1$ neighbours to the right and to the left of each center of a hypersphere are deleted, and saturation in function of w is observed.

The correction is compulsory, even for the longer EEG time series. Results of Grassberger-Procaccia analysis of EEG signals must be considered as having no value when the single-strand correction has not been effected, or if it is claimed not to have affected the results appreciably.

On the other hand, severe warnings about using Grassberger-Procaccia analysis for short time series were given, on fundamental grounds, by the founders of the method [8-10]. The length required for the experimental time series in Grassberger-Procaccia analysis, indeed, increases exponentially with the attractor's dimension d.

The difficulty, however, is not quite as stringent as one might fear, because the problem is not to get the information from single scaled correlation integrals, but from scaled structures involving several embeddings, as we have shown [6,11]. We first

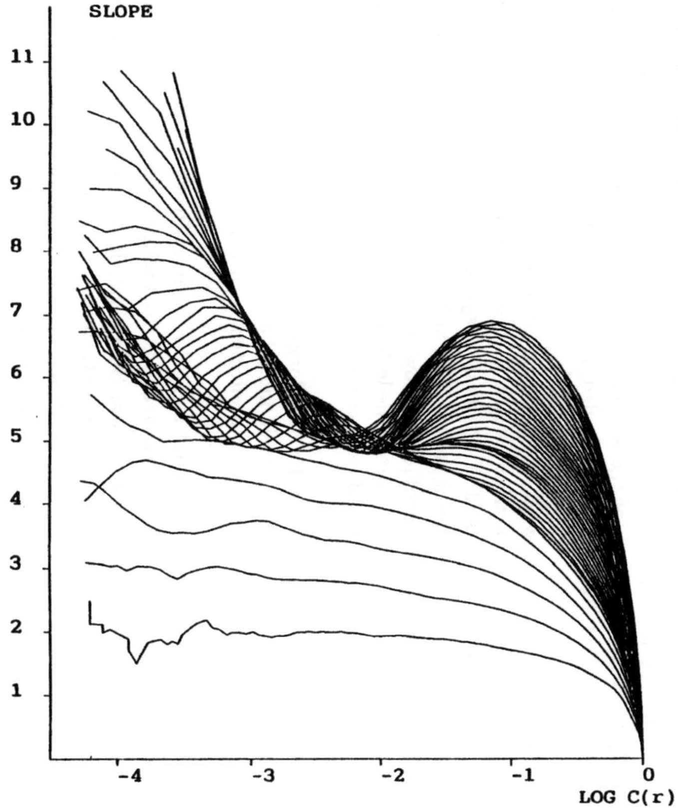

Fig. 1a. Doublet for a solution of Mackey and Glass's differential equation. Slope curves with Theiler's parameter $w = 3$: saturation in function of w is not attained.

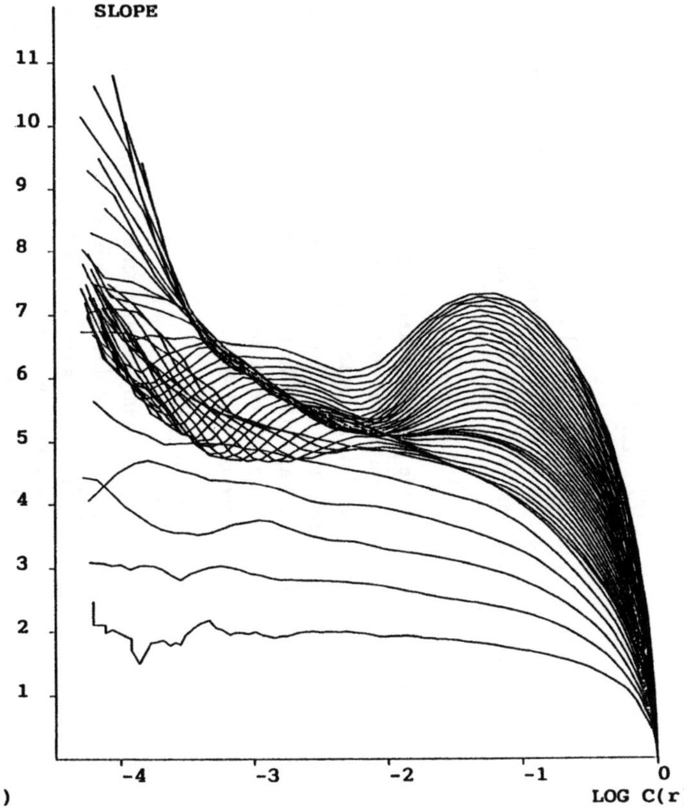

Fig. 1b. Same curves as in Fig. 1a, except $w = 10$: saturation is attained.

observed scaled structures for EEG signals : α-waves, that may be recorded when the subject is in a resting state with the eyes closed, and δ-sleep, that is, deep sleep. Thereafter we also observed similar structures for solutions of the Mackey and Glass differential equation.

In Ref. [6] solutions of the Mackey and Glass differential equation are studied for attractor dimensions close to 5. Examples are given in which, although single correlation integrals show no scaling range, a nearly horizontal power-law structure builds up, that is constructed from different slope curves. This structure appears at the right value of the correlation dimension. As it is contributed by different embeddings, we say that the correlation integrals, or the dynamics, or the system, are "trans-embedding-scaled".

In Figs. 1a and 1b the time series is of only 18.4 pseudo-periods, and the recording frequency is of 136 values per pseudo-period. The number of experimental values is 2,500. The lag time τ in Grassberger-Procaccia's analysis is 4.4×10^{-2} pseudo-period. The pseudo-period is measured from the frequency of the highest contribution to the Fourier power spectrum.

Two structures appear in the figure, the first for $10^{-3} < C(r) < 10^{-2}$, the second for roughly $C(r) < 5 \times 10^{-3}$; the network of slope curves thus exhibits a doublet-structure. In Fig. 1a, $w = 3$; this is the lowest value at which the doublet appears. In Fig. 1b, $w = 10$, a value for which saturation in function of w is attained. Closer scrutiny shows that the right-hand structure is made up of slope curves of the higher embedding dimensions ($n > 16$), and that the left-hand structure is made up of the lower-dimensional slope curves ($n \leq 16$). Remarkably, the low-dimensional slope curves first follow the upper boundary of the high-dimensional right-hand structure, then bend down, each in its turn, to recombine further to the left in the low-dimensional structure. This pattern, which we repeatedly observed for solutions of the Mackey and Glass differential equation and for EEG signals, is radically different from the scaling in individual embeddings for long time sequences.

We called the components of the doublet, both of which are trans-embeddings in the figures, the low-dimensional (LD) and the high-dimensional (HD) components, respectively. However, the embedding dimensions of slope curves that compose a given structure, LD or HD, decrease as the lag time τ in Grassberger-Procaccia analysis is increased, according to the approximate rule $n\tau$ = Cte. It follows that, while τ increases, an HD structure, although it may be rather invariable in position and shape, may turn into an LD structure, when a new HD structure appears to its right, whereas the old LD structure disappears. The old LD structure may also disappear on leaving the former HD structure alone. In some cases, at intermediate values of τ, we even observed three structures simultaneously, that is, a triplet. In these cases, however, the disappearing structure was loosely drawn, so that the pattern essentially remained a doublet.

Fig. 2 is a chart from Ref. [6] summarizing the results which we obtained for Mackey and Glass's equation, when the correlation dimension d is close to 5. In abscissa the recording frequency f is plotted, and in ordinate, the length T of the time series. The crosses indicate conditions for which traditional scaling of single correlation integrals is observed. The hatched area indicates conditions for which the correlation integrals are trans-embedding-scaled. This happens when either the recording frequency is low, or the time series is short. Here the meaning of a long and of a short time series becomes clear. When the time series is more than, say, 50 pseudo-period long, classic Grassberger-Procaccia analysis applies, except at the lower recording frequencies. Below this limit,

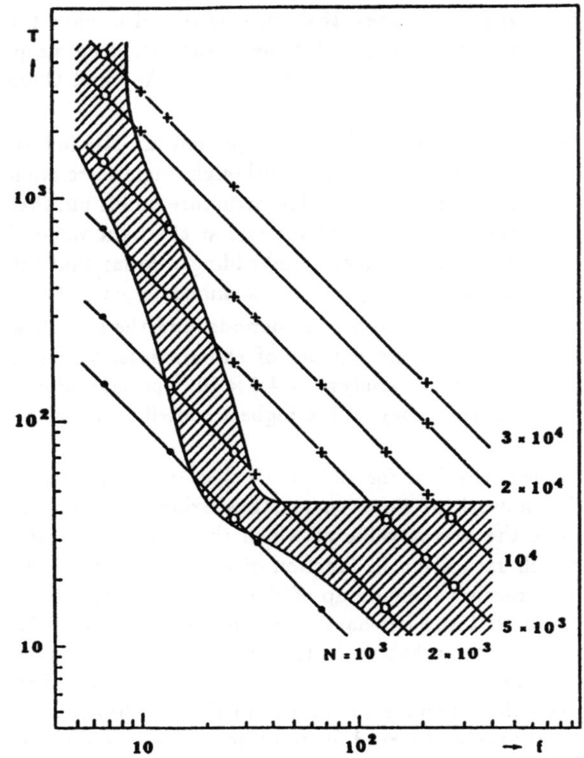

Fig. 2. Chart for scaling regions of dynamical systems of dimension $d \approx 5$, generated from Mackey and Glass's differential equation. Ordinate : length T of time series in pseudo-periods ; abscissa : sampling frequency f in number of data per pseudo-period. Trans-embedding-scaled dynamics occurs in the hatched area (from Ref. [6], with permission).

and above about 20 pseudo-periods in length, that is, exactly in the domain that is essential in EEG analysis, trans-embedding comes to our help. This is also the range in which "doublet-split-scaling" is mostly observed.

In a doublet, the two values of the correlation exponent ν, as measured from the lower boundaries of the LD and HD components, bracket the attractor's correlation dimension d, as determined from the traditional scaling of single slope curves for long data sets. Based on a systematic study of the doublets found for the equation, we defined an ideal doublet, by requiring that the HD value of ν does not exceed the LD value by more than 10 %, and that the LD value is invariant to variation in the lag time τ, to ± 5 %, in a range of values of τ of at least 4 : 1 [11].

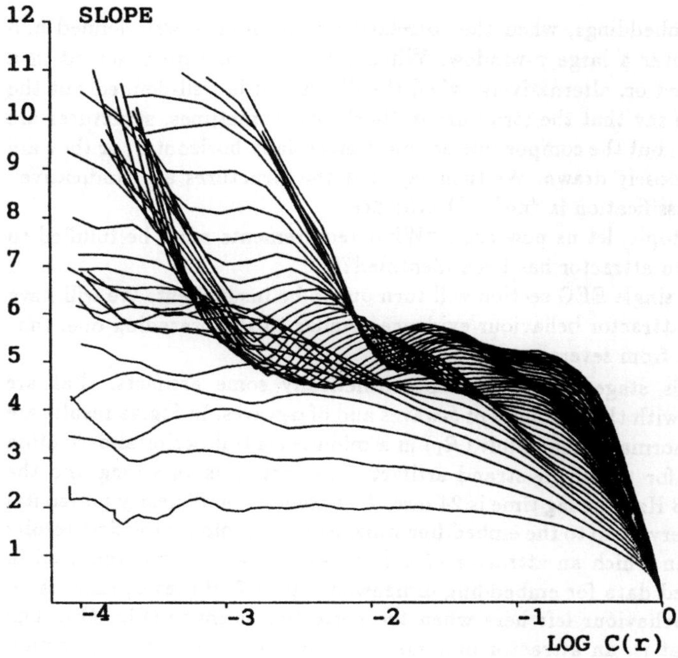

Fig. 3. EEG recording of δ-sleep : the almost ideal doublet of subject R_3 ; 10 s section from a C_zO_z derivation ; f = 128 Hz ; τ = 32 msc ; w = 10 (from Ref. [11], with permission).

3 EEG Attractors : δ-Sleep and α-Waves

In Ref. [11] an example of an "almost ideal" doublet is presented which we found in a δ-sleep recording from a midline central-occipital derivation (C_zO_z) of an insomniac subject (R_3) not undergoing treatment. In Fig. 3 the slope curves are shown for a section 10 s long of the signal, recorded at 128 Hz; $\tau = 32$ msc, $w = 10$; 1 msc = 1 "millisecond corrected" = (1/1.024) ms = 0.9766ms.

In a range of values of τ of 8:1, all values of the attractor's dimension for the LD component were found in the range $d = 4.4 \pm 5\%$. The scaled HD component observed at the lowest values of τ, when τ was increased, progressively turned into a structure pointing downwards. The values of d averaged over the two components of the doublet, when they both scale, or averaged for the LD component and the head of the HD component, when τ is 32 msc or greater, were all in the accepted bracket. Since the sole departure from ideality consisted in the variation in the shape of the HD component, at the higher values of τ, we called this doublet almost ideal.

Next, we had to establish a scale for the quality of the structures shown by an EEG section. The above almost ideal doublet, we classify as "good" (G). We also classify as good lonely trans-embeddings, when the correlation dimension is well-defined and found to be invariant over a large τ-window. When a lonely structure observed in a large τ-window is blurred or, alternatively, when the dimension is well-defined but the τ-window is narrow, we say that the structure is "fair" (F). Sometimes, structures are reminiscent of a doublet, but the components are not tangentially horizontal; or they are nearly horizontal, but loosely drawn. We then say that the structures are "indicative" (I). The last level of classification is "no" (N) evidence.

In the frame of my topic, let us now ask : "What requirements must be fulfilled to allow us to claim that an attractor has been identified?"

Classification of one single EEG section will turn out to be insufficient. We will have to establish a scale for attractor behaviour evidence, similar to the preceding one, that will involve information from several sections.

It is helpful at this stage to illustrate parenthetically some artifacts, that are currently encountered, with the examples of δ-waves and of α-waves. In Fig.4a results are shown for δ-sleep of a normal young adult (R_1) in a midline central-occipital derivation (C_zO_z), not corrected for the single-strand artifact. The section is 14 s long and the recording frequency 128 Hz. The lag time is 24 msc. A concentration of nearly horizontal slope curves can be observed up to the embedding dimension 8. It mimics the first results published in 1985 [1], in which an attractor of a dimension close to 4 was reported in δ-sleep from uncorrected data for embedding dimensions up to 7. However, there is no evidence of attractor behaviour left here when the correction is made (Fig. 4b). The reverse is also true, that is, an attractor may not be detected, because the correction has not been effected.

Similar difficulties are encountered with α-waves, even for longer EEG sections. We have analyzed a 30 s long section of α-waves from a frontal-central derivation (F_3C_3), recorded at 256 Hz. Uncorrected data nicely showed attractor behaviour of a dimension slightly below 5. No attractor behaviour was left in the corrected data.

It is highly instructive also to shift a section along the time axis. Thus, for example, the 10 s long δ-sleep section of subject R_1, that yielded the beautiful scaling range of Fig.11 in Ref. [11], when set 1 s earlier revealed only traces of attractor behaviour, of the kind we call indicative.

We call an ensemble of sections obtained by shifting a section and/or changing its length, the sections "around" the one we started with. We also call this ensemble a "segment" of our EEG signal.

For the δ-sleep section of subject R_3, that yielded our almost ideal doublet, the results were as follows : among 4 sections starting at the same instant, the 3 where the length did not exceed 14 s were good. Among 4 sections 10 s long that were shifted along the time axis, the 3 where the shift did not exceed 4 s, were also good. The 2 non-good sections were fair. In this case we say that, among the ensemble of sections, that is to say for the segment, the evidence for attractor behaviour is "good".

We can define in a similar way segments giving "fair" and "indicative" evidence, as well as "no" evidence, and, thus, the same terminology we use for classifying the scaled structures in a section can also be applied to the evidence of attractor behaviour in a segment.

Table I summarizes the results for δ-sleep in our subjects R_1, R_2, and R_3 of Ref. [11], in two derivations. For each subject and derivation the number of results corresponding to each classification is given. The total number of segments analyzed per subject and

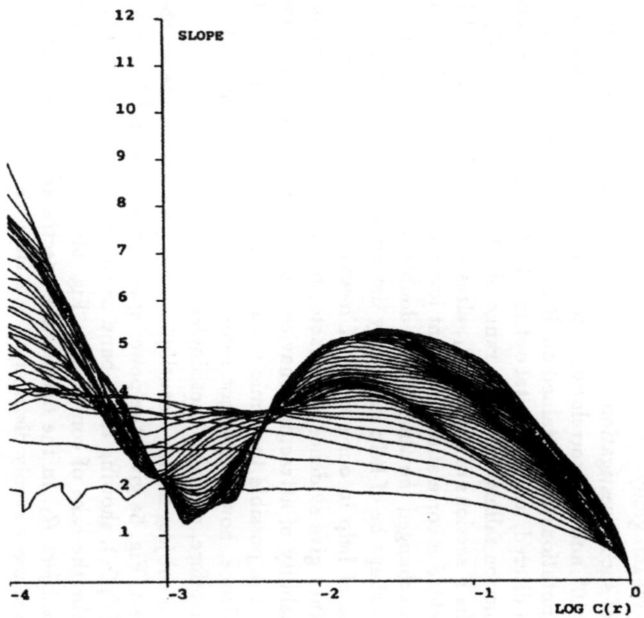

Fig. 4a. EEG recording of δ-sleep (subject R_1); 14 s section from a C_zO_z derivation ; $f = 128$ Hz ; $\tau = 24$ msc. No Theiler correction ($w = 1$). For embedding dimensions $5 \leq n \leq 8$ and $10^{-3} < C(r) < 10^{-2}$ the slope curves mimic the traditional scaling of correlation integrals, as well as attractor behaviour $d \approx 4$.

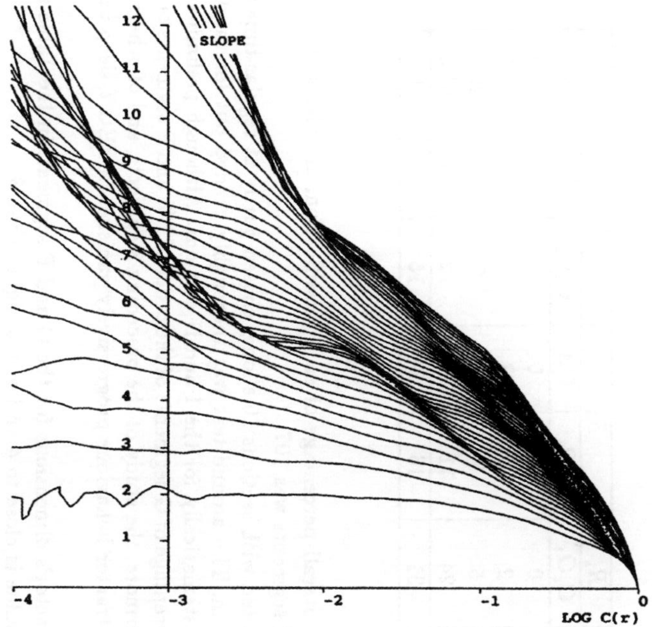

Fig. 4b. Same curves as in Fig. 4a, but Theiler-corrected ($w = 10$). No evidence of attractor behaviour is left.

Table 1. Classification of EEG "segments" showing attractor behaviour, and their average probabilities (last column) : three subjects and two derivations.

	R_1	R_2		R_3		
	C_zO_z	C_zO_z	CA	C_zO_z	CA	%
G	0	0	0	1	0	0.9
F	2	2	3	0	1	7.5
I	5	3	13	2	7	28
N	24	11	8	12	13	63.6
107	31	16	24	15	21	100

derivation, and the overall percentage of each classification, are also given. The total number of analyzed segments was 107.

The analysis started with sections 10 s long, chosen partly at regular time intervals and partly at random. The around-test was thoroughly carried out for the G and F sections, but not systematically for the I and N sections. Although preliminary, these results suggest a percentage of G segments of the order of 1 % and F segments of between 5 and 10 %. Furthermore, the study of the preceding 3 subjects, and of others, showed that evidence for attractor behaviour progressively vanishes in EEG δ-sleep recordings of more than 12 s.

Most of the attractor's dimensions for the G and F segments in the Table are close to 5, although one value is close to 4 and one is higher than 6.

It will probably be possible in systematic investigations to use not only G segments, but also F segments meaningfully.

The completion of such investigations for a population of 10 subjects would require computing between 10^8 and 10^9 correlation integrals. Useful hints, that can perhaps help reduce the computations, are obtained on filtering the signals. Some of our δ-sleep signals were low-pass filtered, using a digital optimal FIR (finite impulse response) filter, with the transfer function falling off in the range of 20–30 Hz. As a result, the scaled structure in one of the F sections of subject R_1, already referred to (Fig.11 of Ref. [11]), deteriorated, and yielded a correlation exponent lower by 1 unit. On the contrary, the same procedure left unchanged the almost ideal doublet described above for subject R_3. Thus, filtering will perhaps be of help in quickly detecting good segments.

Filtering was also of help in our studies of α-waves. The trouble with α-waves is that very rarely do they give evidence of attractor behaviour (one example is found in Ref. [12]). The probability of attractor behaviour is presumably less than for δ-sleep. On the other hand, it is possible to mimic attractor behaviour on pass-band filtering an α-band. The evidence, however, is not robust to small alterations of the filtering procedure, and is, therefore, probably meaningless.

Nevertheless, for a subject having beautiful α-waves, we have made an observation that may be of interest. Fig. 5a gives the power spectrum for this subject in a parietal-occipital derivation (P_1O_1), showing the α-band as well as components of lower and of higher frequencies. For the sake of comparison, Fig. 5b gives the spectrum of δ-sleep for a section of our subject R_3. In the figures, the spectra are square-rooted. Spectral contributions are much more evenly distributed in δ-sleep than in α-waves.

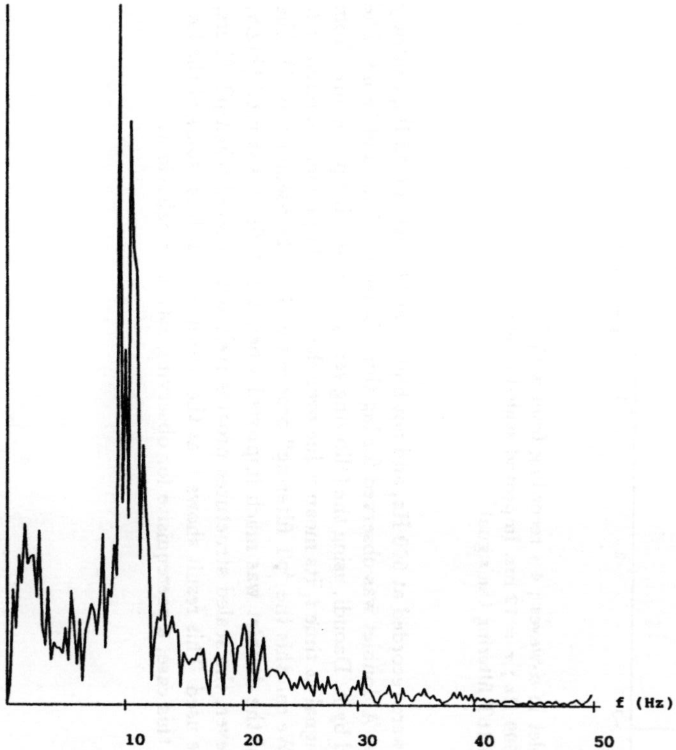

Fig. 5a. Fourier power spectrum of α-waves from a $P_1 O_1$ derivation. The signal was 16T filtered.

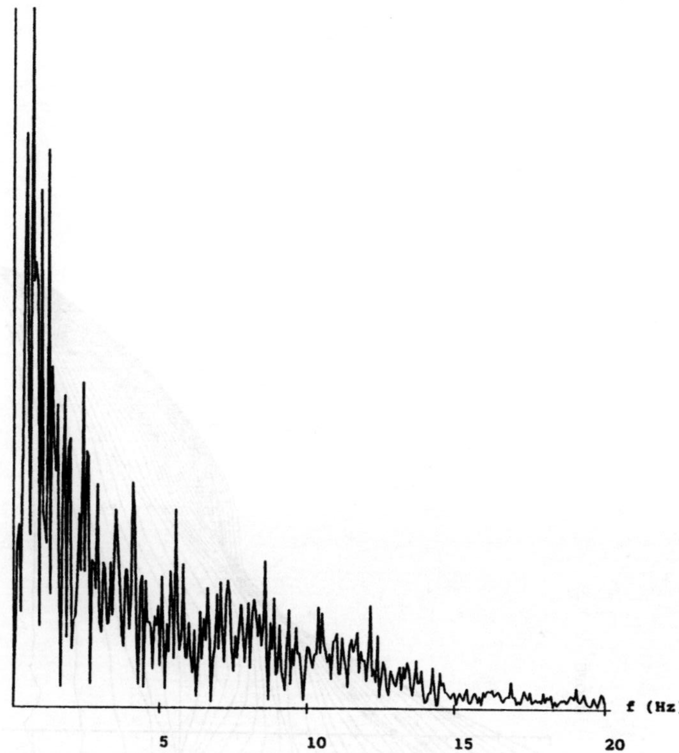

Fig. 5b. Fourier power spectrum of δ-sleep for the almost ideal doublet of subject R_3 (cf. Fig. 3).

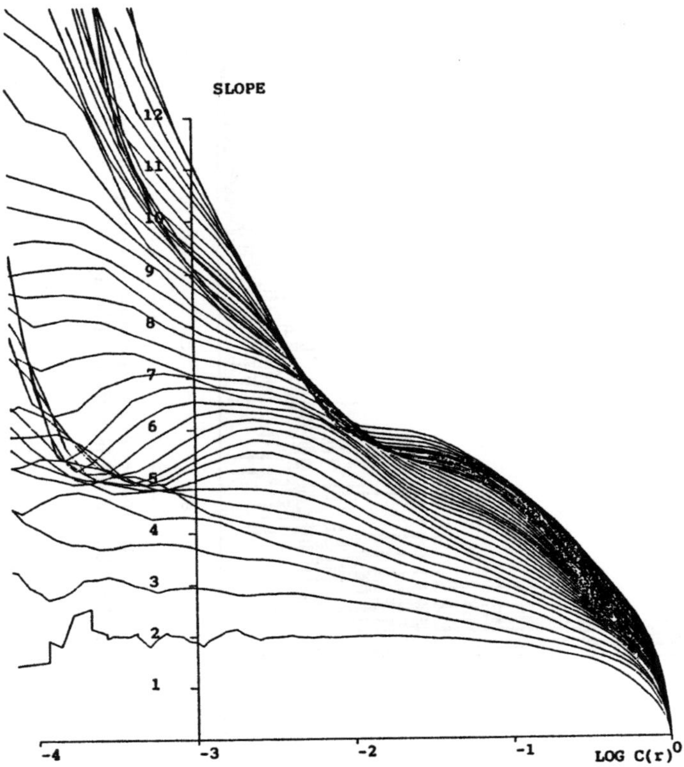

Fig. 6. Doublet in α-waves ; 4 s recording from a P_1O_1 derivation ; $f = 500$ Hz ; $\tau = 12$ ms. Improved scaled structures obtained on 16T filtering the signal.

The α-waves were recorded at 500 Hz, and not high-pass filtered at 0.53 Hz, contrary to clinical routine. A doublet was observed for lag times between 10 ms and 18 ms. The signal was treated by A. Daoudi, using the following very gentle filtering procedure. From the value of the signal at time t_i its mean-value over the p pseudo-periods centered at t_i was subtracted. We call this the "pT filtering" procedure (T = the pseudo-period). The LD component of the doublet was much improved when a 16T filter was used (Fig.6; $\tau = 12\,\text{ms}$); however, the scaled structures progressively deteriorated when pT filters, with $p \leq 12$, were used. This result shows that the presence of at least some of the low frequencies is, in this case, a prerequisite for observing attractor behaviour.

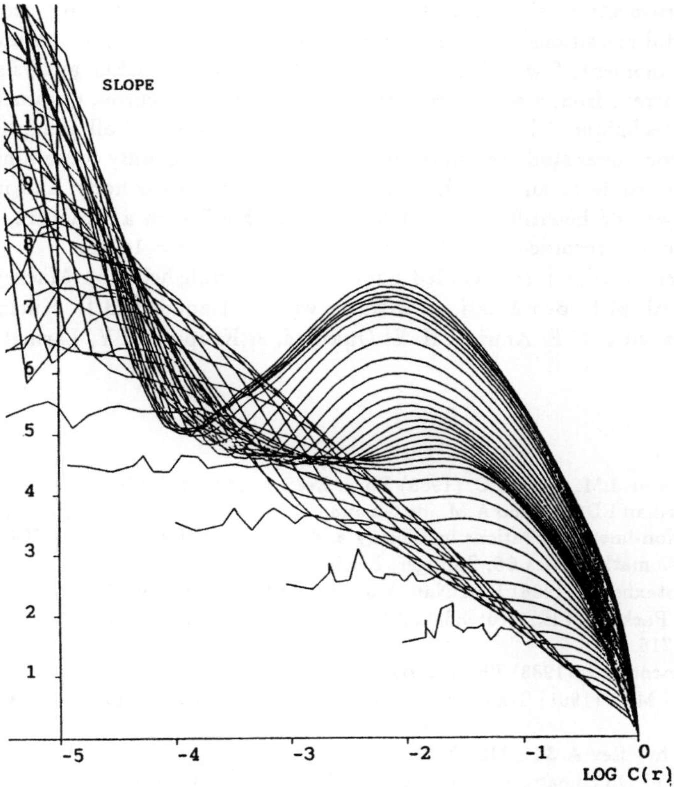

Fig. 7. Trans-embedding in 7 s recording of an ionic current from a single preganglionic sympathetic neuron (patch clamp technique) ; $f = 1024$ Hz ; $\tau = 14$ msc ; $w = 9$.

4 Conclusions and Outlook

If we had not found our unique good segment, I would not have dared to give a positive answer to my questioning about EEG attractors. This good segment, however, is for us a kind of standard, that permits us to believe that evidence of lesser quality may cover useful experimental data.

Our description, in any case, differs profoundly from previous ones. We find attractors essentially in δ-sleep, with life times that barely exceed 10 s, and with low probabilities. Furthermore, our preliminary results for α-waves suggest that, perhaps, attractors in these waves also require the presence of low frequency components.

Although it is too early as yet to see a picture emerge, we can speculate that slow waves, which are concomitant with increased coherence of brain activity, are a prerequisite for attractor behaviour.

What about the prospects of these investigations? Perhaps we now mainly need controllable experimental conditions in which attractors are observed with not too low a probability. For the moment, I would like to cast a glance beyond EEG analysis. We studied the ionic current from a single preganglionic sympathetic neuron, obtained by the "patch clamp" technique. The recorded current is the resultant of all synaptic currents which the neuron under study registers. In spite of its tracing, mainly containing brief impulses, it was possible to analyze this signal with the above methods, and in some of its sections it yielded beautiful trans-embeddings. In Fig.7 such a structure is shown for a 7 s long section, recorded at 1024 Hz; the lag time was $\tau = 14$ msc.

The δ-sleep data are in print [11]. Detailed papers will be published : on Mackey and Glass's equation with M.L. Ben Maati, on α-waves with A. Daoudi and D. Kurtz, and on neuronal signals with M. El Amri, E.H. El Ouasdad, J. Krupp and Y. Larmet.

References

1 Babloyantz A., Salazar J.M., Nicolis C. (1985) Phys. Lett. A **111** : 152-156

2 Rapp P.E., Zimmerman I.D., Albano A.M., de Guzman G.C., Greenbaum N.N., Bashore T.R. (1986), in : Non-linear oscillations in Biology and Chemistry. Edit. Othmer H G, Lecture Notes in Biomathematics **66**, Springer, Berlin : 366-381

3 Babloyantz A., Destexhe A. (1986) Proc Natl Acad Sci USA **83** : 3513-3517

4 Ruelle D., cited in Packard N.H., Crutchfield J.P., Farmer J.D., Shaw R.S. (1980) Phys Rev Lett **45** : 712-716

5 Grassberger P., Procaccia I. (1983) Physica **9D** : 189-208

6 Cerf R., Ben Maati M.L. (1991) Trans-embedding-scaled dynamics. Phys Lett A **158** : 119-125

7 Theiler J. (1986) Phys Rev A **34** : 2427-2432

8 Grassberger P. (1986) Do climatic attractors exist? Nature **323** : 609-612

9 Procaccia I. (1988) Weather systems. Complex or just complicated? Nature **333**: 498-499

10 Ruelle D. (1990) Deterministic chaos : the science and the fiction. Proc Roy Soc., London Ser A **427** : 241-248

11 Cerf R., Oumarakchi M., Ben Maati M.L., Sefrioui M. Doublet-split-scaling of correlation integrals in non-linear dynamics and in neurobiology. Biol. Cybernetics, in print.

12 Cerf R., Daoudi A., Khatory A., Oumarakchi M., Khaider M., Trio J.M., Kurtz D. (1990) Dynamique cérébrale et chaos déterministe. C R Acad Sci Paris **311** II : 1037-1044

Synergetics of Blood Movement Through Microvascular Networks: Causes and Consequences of Nonlinear Pressure-Flow Relationships

H. Schmid-Schönbein

Institut für Physiologie, Klinikum der RWTH Aachen,
Pauwelsstr. 30, D-5100 Aachen, Fed. Rep. of Germany

1. Introduction: From Ludwig von Bertalanffy to Hermann Haken

1.1 Extension of physiological systems analysis

The application of Haken's concepts of synergetics to biological systems analysis is on its way [1-3]: on the macroscopic scale, the observation of non-equilibirum phase transitions in progressively driven systems testifies to the general applicability of his coherent interactionist theory. Multiple mutual functional links between various forms of energy conversion processes with high efficiency (see below) make the operation of synergetic principles very likely. However, skeptical and reductionist biologists are not necessarily a priori impressed by the derivation of biological synergetics from first physical principles (forcing them to follow a "top down analysis" which is not always transparent to them). The host of novel experimental tests in macroscopic biological system analysis cannot replace a much more thorough elaboration of the underlying mechanisms on the mesoscopic and on the microscopic level[1] : in short, we need the bottom up approach.

Those of us impressed by Haken's bold generalizations are mesmerized by two aspects of "general synergetics", namely that
1) homologous regularities should govern the complex behaviour of ensembles of molecules, cells, organs, organisms (and societies of organisms) and that
2) paradigms developed in thermodynamics, a science in command of the common "language" of all natural sciences, should provide the logic allowing to put each individual mechanisms, and - in the long run - the whole of Haken's visionary concept to critical testing.

This was done so far by the analysis of putatively "synergetic" (and not just synergistic) modes of operation of macroscopic systems by monitoring (often over extended periods of times) their dynamic fluctuations (see the contribution of Koepchen in this volume).

This notwithstanding, research in physiological synergetics at the mesocopic and the microscopic level is a challenge for the future; here , this new paradigmatic system must be put into a common perspective with traditional biological concepts. In order to achieve this, synergetic systems analysis should, whenever possible, take up the semantics and conventional classification criteria

1 It is perhaps appropriate for a synergetic "insider" to warn against the danger of self-fulfilling prophecies, especially since synergetics - in its rational explanation for circular causality due to the enslaving principle might prove to be a seductive but slippery road leading to circular logic

Table 1: List of interacting Bertalanffy - Systems found in "Vegetative" Physiology: systems primarily devoted to the transfer of matter and energy (plus transfer of chemical information)

1. INTRACELLULAR COMPARTMENTS ("boosted" by ionic pumps)

1.1. the individual intracellular spaces
1.2. the global intracellular compartment

2. EXTRACELLULAR COMPARTMENTS ("boosted" by the heart)

2.1. the local extracellular space
2.2. the global extracellular space
 2.2.1 the interstitial space (local, global, containing the lymph-systems)
 2.2.2 the paracellular water spaces (intestine, tubular fluid, ocular fluid, various intracerebral fluid spaces).
 2.2.3 the intravascular space
  (i) high pressure system (storing energy)
  (ii) capacitance system (low pressure system,storing material)
  (iii) the microvascular exchange systems in the different organs (dissipating energy and material)

for delineating distinct functional systems[2] and the very sound biological rationalizations justifying such distinctions. Here, one can start where Ludwig von Bertalanffy [4] ended and define in his words a "steady state system" as open and adaptive and perceive it as being simultaneously in progressive composition and decomposition. Von Bertalanffy anticipated three fundamental principles of biological synergetics, namely that (1) system structure emerges during function, (2) structure and function are inseparably interwoven, and (3) certain "attractor processes" become only transiently manifest.

It is fascinating to note that for all of the subsystems (Table 1) physiologists have learned to distinguish since Claude Bernard (and which all have unique ways of transfering energy, matter and information) are in fact governed functionally, and of course energetically - by very similar regularities.

It is perhaps trivial to state that the hierarchically listed subsystems distinguished in the domain of vegetative physiology are all characterized by their being permanently "driven" to a non-equilibrium state by the action of the heart as a pump. In the terms of Bertalanffy, even when at rest they are in a "flow equilibrium" (Fließgleichgewicht) or "boosted steady state" in that a major part of the mechanical energy generated in the heart is (more or less permanently) stored in them as potential energy, kinetic energy or "colloid-osmotic energy". Physiological activity is initially driven by the passive flow of energy "penned" in the boosted compartments, which is subsequently replenished by increased energy conversion processes and import of free energy. Thence one can say that at the boundaries of driven systems in the sense of Table I there are highly potent control parameters (very high generalized forces) which stabilize energy transfer processes (very stable generalized flows).

2 The greek root of the terms "system" nature (from nascere, to be born) and contain much more of the conceptional mainstream of Haken's synergetics than often realized. The word "system" is contracted from syn "histanai" to "synistinai" and then systema. Interestingly, the modern indo-german languages have not developed a simple word allowing a direct translation of "synhistinai", which means something like "to cause to stand together, German "gemeinsam entstehen". The term "emergence", German "Auftauchen", (derived from latin "mergere" = to dip, to plunge) is interestingly defined in Webster's dictionary by "rise from an enveloping fluid".

Since the transfer processes for energy and matter utilize coherently operating conducting channels or pathways which one might call "generalized channels", it can be said that the very high generalized forces in themselves lead to highly coordinated movement of energy and matter. However, as will be shown below, the control parameters cannot simply induce movement, but can enhance mobility of moving particles (for example erythrocytes) thereby causing a mesoscopic second order phase transition which produces a highly non-linear pressure-flow relationship.

One might look for similar systems in "Animalic" physiology serving the transfer of information (utilizing highly localized and strictly coherent flow of energy and matter across "synapses") and induction of locomotion. These could be subclassified into

1. The senso-motor systems responsible for the selective import of "informational energy", and for orderly and coherent movement of the body (or parts of the body),
2. the central and peripheral nervous systems: distributing and processing informational flows in large neuronal networks by highly coherent and temporally ordered transfer of energy and matter and lastly
3. the moto-sensory systems: serving programmed and selforganized transfer of chemical into mechanical energy under the control of self-organized neuronal pools.

1.2 Definition of "Boosted Steady States"

Physiological systems analysis can start with the generalization that the resting state is characterized by very high metabolic activity, which in boosting a steady state generates substantial amounts of entropy. As a "side effect", very high "generalized forces" are put on hold, including extremely steep chemical (including osmotic), electrical, electrochemical gradients, as well as pressure gradients, shear stress gradients, gradients of kinetic energy and, of course, local gradients of affinity. In addition, the "boosting" structures or tensilizes the configuration of channels for the export of energy and matter across the boundaries of the systems described above.

The generalized channels"in this sense include ligand and voltage gated membrane pores, and intramural pores in vessels. As demonstrated later, even the smallest blood vessels connecting arteries and veins can be included in this category. The "generalized channels", in limiting the export of energy, contribute to and control position of the "non-equilibrium state" at rest.

Thence, they codetermine the magnitude of the "generalized forces" acting at the system boundary and, in their capacity to structure the configuration of the limiting conductances in the resulting "generalized flows". This mode of "dualistic organization" (high force, precise pathway) allows explosive increases in the transfer rate of energy and matter across the system boundaries. Therefore, with the aid to sumpplement the old definitions given by von Bertalanffy [4], namely "steady state" and flow equilibrium", the present author has previously [5] coined the term "boosted steady state" to describe the result of a dynamic evolution of highly ordered physiological systems where displacement far away from the thermodynamic equilibrium has dual effects on both structure and function. This concepts allows to include the contemporary knowledge about the microscopic and mesoscopic structuralization of macromolecular system components, a dynamic process which emerges from the continuous stream of energy and matter. What is meant is a hybrid of conservative structure and dissipative structuring and it is proposed to call it "quasi-conservative structuring".

A few examples shall suffice to illustrate the "boosting" principle detailed elsewhere [6].

It applies to excitable membranes: here, this means it that the conformation of protein channels at rest is incessantly influenced by the membrane charge (and/or the subthreshold membrane current) that are continuously flowing across the boundaries of systems (through leaks, pores or partially active channels, see Hille [7]). In bio-fluid-dynamics, a similar situation is created in the so-called arterial windkessel [8,9] for the high pressure system), where the arterioles, in limiting the export of energy and matter, on the one hand help in boosting the upstream arteries filled by the pumping left ventricle (see [8]). The built-up pressure, on the other hand, via its influence on the arteriolar wall tension, leads to the constriction of arterioles at rest, and thence to very low output conductance (see Section 3.3).

The term "boosted steady state" can useful describe the functional resting state as a situation where the "generalized forces" and the "generalized pores" produce not just a stable non-equilibrium state close to a kinetic threshold (and thence "criticality"), but a situation where at rest molecular and cellular structures are "tensilized" by continuous energetic expenditure in such a fashion that they limit in an active fashion the export of energy and matter. In open driven systems, therefore, energy and matter can be said to be actively penned, ready to erupt after release of tension into more or less coherent movement during various kinds of physiological "activity". It is this energetic mode of operation that introduces extreme non-linearity into functional operation of "boosted systems". The "trigger mechanisms" initiating physiological activity quite often operate by simply removing the dynamically induced limitation to a conductance after minute stimuli by a passive process of muscle relaxation. Thereby, a sudden increase in the transfer of energy and matter may occur (a reaction which phenomenologically can produce extremely non-linear force-effect relationsship or "soft stimulation "(weiche Anregung) a term coined by Grossmann [10]) into physiological activity.

The maintenance of "boosted steady state" requires a very high and continuous energetic expenditure (measurable as the basic metabolic rate of roughly 100 Watt in a young man of 70 kg) and thence for very high entropy generation rates at physiological rest. This is the price paid for the ability to explode into physical activity and the potential for rapid increments in the transfer and conversion rates of energy and matter. We can further define "synergetic operation" in the general sense that a biological system exhibits spontaneous transitions into more ordered forms of cooperativity, which, however, emerges only in physiological activity, i.e. when energy conversion processes are "accelerated" (but not when of rest). Examples for this transition into synergetic cooperativity of electrical, chemical and mechanical events can be found during muscle contraction (see [6]). It is proposed that transition into synergetic co-operation in the strict (thermodynamic) sense of the word is accepted, provided that after crossing a kinetic threshold the criterion of decreasing excess entropy generation (Klimonovich and Ebeling [11]) can be fulfilled. This criterion means that with increasing rate of energy transfer from the resting to the active state either the absolute, or at least the relative increase in the rate of entropy production in the system is decreased, this being the quantitative thermodynamic proof for the emergence of a more coherent mode of the system of operation.

At this point, unique "kinematic" mechanisms come into the play, namely dynamically induced second order phase transitions taking place far from thermo-dynamic and fluid-dynamic equilibrium (see below).

1.3. "Dissipative Structuring" and "Dynamic Rectification" of Particles Subjected to Thermodynamic Forces

Driven biological systems are composite materials, i.e. solutions of ions in water, dispersions of macromolecules in fluids, of cells in solutions of macromolecules. When subjected to very high thermodynamic forces, the phase relationship between the continuous phases and the dispersed phases can be changed, simply because these relationsships (for example the strengths of hydration of ions) are "susceptible" to the thermodynamic forces which the ions can become exposed to. A similar phenomenon is observed in the behavior of mammalian erythrocytes, which are mechanically susceptible to the pressure and shear stress gradients that act on them when they are moving in blood vessels.

This report focusses on the kinematics of the latter phenomenon in non-equilibrium b i o - f l u i d dynamics (namely blood flow) here presented as a typical process of "dissipative structuring". In allowing the movement of a complex multiphase fluid (containing particles with 8 μm diameter) through microscopic vessels (diameter 5 μm), a physically ordered movement of matter not unlike that observed of electrons in superconductors takes place, effectuated in response to the physical forces acting on the dispersed phase subjected to strong hydrodynamic potentials. It is characterized by the emergence of responses minimizing entropy generation rates in a system operating far from fluid-dynamic equilibrium. In essence, a new type of interference effect in a binary system is described, where in a moving multiphase fluid increasing shear stress gradients produce progressive concentration gradients which in turn produces a preferred mode of operation (or a "stable instability" in the conventional terminology of irreversible non-linear thermodynamics) [12]. In the preferred mode of operation, the bilinear dependence of entropy generation is locally conserved, but globally violated, and the two phases are no longer moving at identical pace, producing a break (breach) of symmetry (Bruch im Gleichmaß der Bewegung) at a new dynamic boundary layer where different order parameters (here: local Reynold's numbers) are evolving. The result of the "local rectification" of the individual moving erythrocyte not just increases its thermodynamic mobility, but "globally induces" coherence of the motion of blood cell suspensions in large network of microvessels; provided that all of the cells in all of the network segments are rectified in a coherent fashion. This, of course, requires the maintenance of the "boosted steady state" in the entire macroscopic system. I will demonstrate its highly effective operation in the boosted steady state, but also its breakdown near equilibrium after "critical slowing" has occurred and one can recognize the extent of physiological order by contrasting it to the "chaotic behaviour" of a complex system near equilibrium.

2. Synergetics of RBC-movements

2.1 The Rheology of Rectified Movement of Red Blood Cells in Microvessels

The biomechanical event of coherent perfusion of microvascular networks is closely associated with the very high fluidity of the highly concentrated red blood cell suspensions as a consequence of a highly specialized adaptation of the individual red cells (the oxygen carriers) to the forces acting upon them in flow. This process shall be called "erythrocyte rectification" [13]. The mechanical problem of this movement becomes immediately evident when one considers the geometric boundary conditions: erythrocytes amount to roughly 45% of the volume of blood, their diameter as resting biconcave "discs" is roughly 7.5 μm, the diameter of the nutritive capillaries is roughly 5 μm. When driven by the physiologically acting pressure gradients along the capillaries (roughly 1000

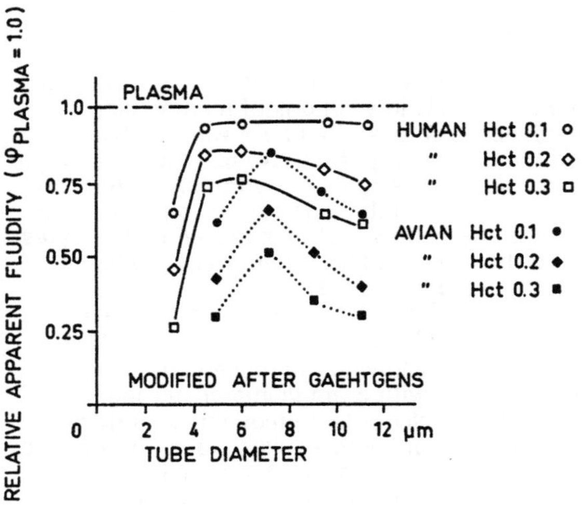

Fig. 1. Relative apparent fluidity of suspensions of flexible and highly "fluid-drop-like" human erythrocytes plotted as a function of volume fraction and tube diameter (open symbols and connected lines). Note that relative apparent fluidity shows pronounced maximum close to 1.0 in tube diameters down to 4.0 μm. Data for less deformable and nucleated avian red blood cells (closed symbols plotted lines are shown for comparison)

Pa/mm), however, the cells are continuously deformed into a "bullet shaped" configuration (diameter roughly 3.5 μm) they assume a stationary orientation, and - by a unique mechanism they - participate in flow due to the continuous rotation of the erythrocyte membrane around the fluid cytoplasm. The details of these combined mechanism paraphrased as "shear induced erythrocyte elongation " have been repeatedly reviewed (see Schmid-Schönbein [13-16]): its kinematics will here be briefly discussed in a synergetic context as a self-organized process which results in a dramatic reduction of entropy generation associated with flow.

The kinematic phenomenology of the latter effect is long known as the formation of a clear plasma layer near the wall of capillaries perfused by red cell plasma mixtures (called "couche de Poiseuille in the french literature). In hemorheology e.g. [18], this is known to cause the so called "Fahraeus-Lindqvist-effect", i.e. the progressive increase in apparent blood fluidity as blood moves from the larger vessels (3-1 cm diameter, apparent viscosity roughly 4 mPa s at 37° C), to the microscopic vessels (5 μm diameter, apparent viscosity roughly 1.2 mPa s or only 70% higher than that of H_2O). Moreover, contrary to the prediction of Einstein-Stoke equation, the apparent viscosity of the blood in these microscopic vessels perfused with steep pressure gradients is largely independent of the volume fraction of the dispersed phase (Fig.1).

Thus, when subjected to the steep pressure gradients found in microvascular beds (which act as the (main principal control parameter) and the steep gradients of shear stresses (which operate as additional attractors), the individual red cell undergoes a transition from a quasi-solid to a quasi fluid body, inducing a transition of red cell dispersions from a suspension of solid particles to an emulsion of functionally fluid "droplets". Therefore, in the absence of the physiologically steep pressure gradients (i.e. when the blood moves near fluid dynamic equilibirum), the same material in a reversible fashion exhibits a very low apparent fluidity (Fig.2) without undergoing coagulation [20].

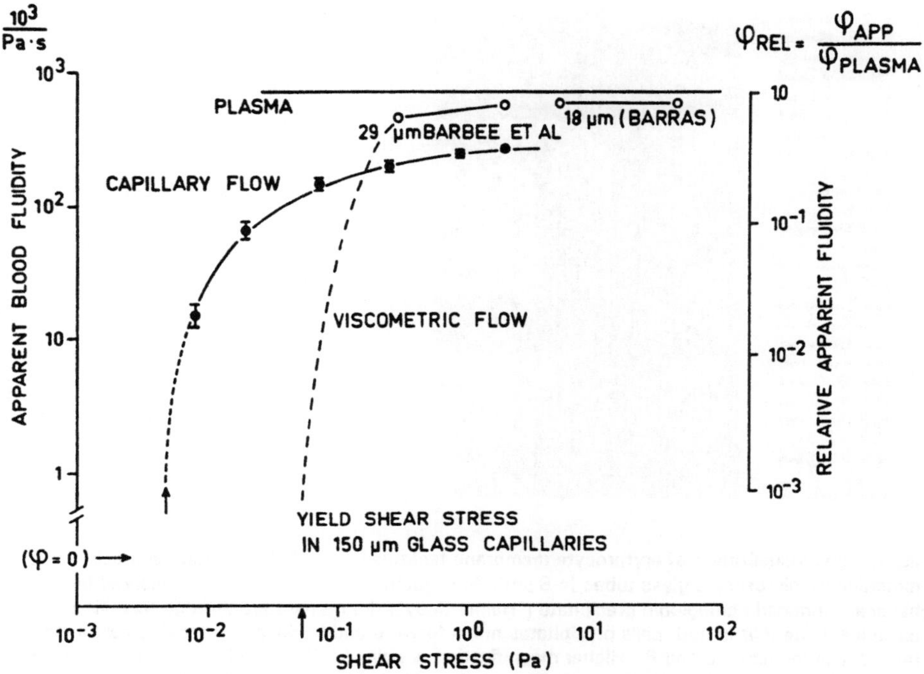

Fig. 2. Relative apparent viscosity of normal human blood plotted (volume fraction 0.4) as a function of incident shear stress and comparing data obtained in capillary flow and viscometric flow. Note high values of relative apparent fluidities at high shear stresses (but strong decrease in fluidity at low shear stresses near fluid-dynamic equilibrium.

2.2 Kinematics and Energetics of Second-Order Phase Transition of Human RBC in Viscometric Flow in Vitro

A bewildering confusion of terms has been used in biophysics, physiology and hemorheology to describe the simple fact that a dynamic second order phase transition takes place in suspensions of non-nucleated mammalian erythrocytes when subjected to high shear stresses. Thus, the concepts and the terminology developed in non-equilibrium thermodynamics and statistical mechanics not only bring the observed physical behaviour into a coherent kinematic perspective, but produce semantic economy. The pivotal process underlying the "rectification" of moving erythrocytes (the transmission of shear stresses from the suspending phase into the inner of the moving erythrocyte by the mechanism of membrane "tank-treading" (see Fig.3) bears striking resemblance to classical self-organized fluid-dynamic processes such as the formation of "Benard rolls" and Taylor vortices (see Chandrasekhar) > " [21].

It goes without saying that many of the kinematic details of this adaptation to a complex flow field remain to be elucidated: suffice it to say here that due to the fluid nature of the cytosol (a concentrated hemoglobin solution devoid of cell organelles or other forms of a dispersed phase) and due to the visco-elastic behaviour of the cell membrane (a two dimensional, incompressible fluid with

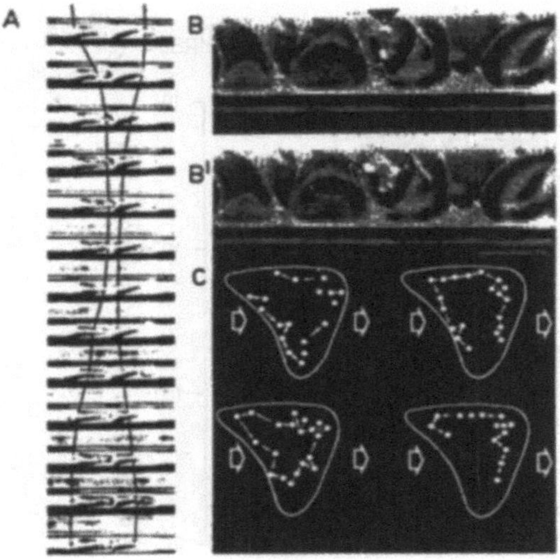

Fig. 3. Flow visualization of erythrocyte membrane tanktreading, cell deformation and cell elongation in microscopic glass tubes (4-8 μm). A: sequential tracings of red cells marked by chemically induced hemoglobin precipitate ("Heintz-Body") during passage through travelling capillaries. Note that in both cells precipitates move forward and backward in moving cells with stationary outershape. B and B': Higher magnification: a pair of "Heintz-Bodies" turning while cell moves without change in outershape. C: Trajectories of individual "Heintz-Bodies" while red cells remain stationary in outershape but move down capillary.
(Scenes taken from a film produced in cooperation with Prof. Peter Gaehtgens, Berlin and shown during conference)

highly non-linear mechanical properties (see [22]) a very special type of "dissipative structuring" is initiated. This is not just of obvious interest for one of the most fundamental transport processes in biology (see [13]) but of considerable theoretical interest in non-linear non-equilibium fluid dynamics (and thermodynamics). This follows from the fact that far from equilibrium one observes a dual and of course self-stabilizing process leading to progressively decreasing entropy generation.

In order to appreciate the functional and the thermodynamic benefit of this mode of operation, one has to keep in mind that reversible binding of oxygen requires the presence of high molecular weight hemochromes, the transport of which is bound to increase the entropy generation associated with convective transport (e.g. Poiseuille flow).

Due to the stationary orientation of the individual erythrocyte, the additional entropy generation associated with the movement of the individual erythrocyte along with the plasma is approaching its theoretical minimum (zero); likewise, due to the passive drift of all erythrocytes from the marginal layers to the axial ones, the overall entropy generation rate is mimized, i.e. reduced very near to the smallest conceivable value (Fig.4).

Both on the local level and on the global level, the transport process of hemoglobin by way of the non-nucleated erythrocytes (which were a phylogenetically late achievement obtained by the genus of the marsupials and the mammals) [23] is almost free of "energetic cost". This follows from the well

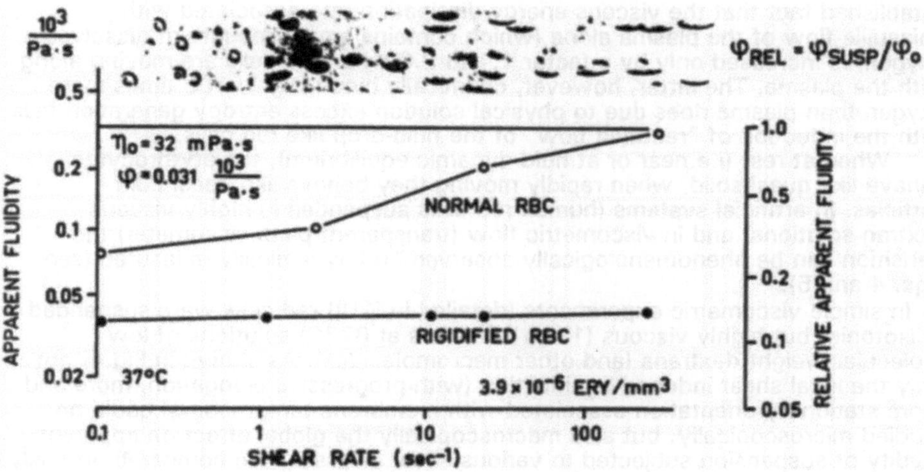

Fig. 4. Model experiments of normal and hardened erythrocytes subjected to viscometric flow (coaxial cylinder rheometer) and suspended in highly viscous isotonic Dextran solutions. Absolute (left ordinate) and relative (right ordinate) fluidity plotted as a function of incident shear rate at adjusted erythrocyte count (3.9 x 10^6 RBC / mm^3). Photomicrograph of erythrocytes taken in transparent cone plate flow chamber ("rheoscope") at various rates of shear are also shown: cells discoidal in shape at low values of are deformed into prolate ellipsoids. As the shear stresses are increased, the cells become progressively "rectified", i.e. elongated, oriented and their membrane rotates around the fluid cytosol (see Fig. 5).

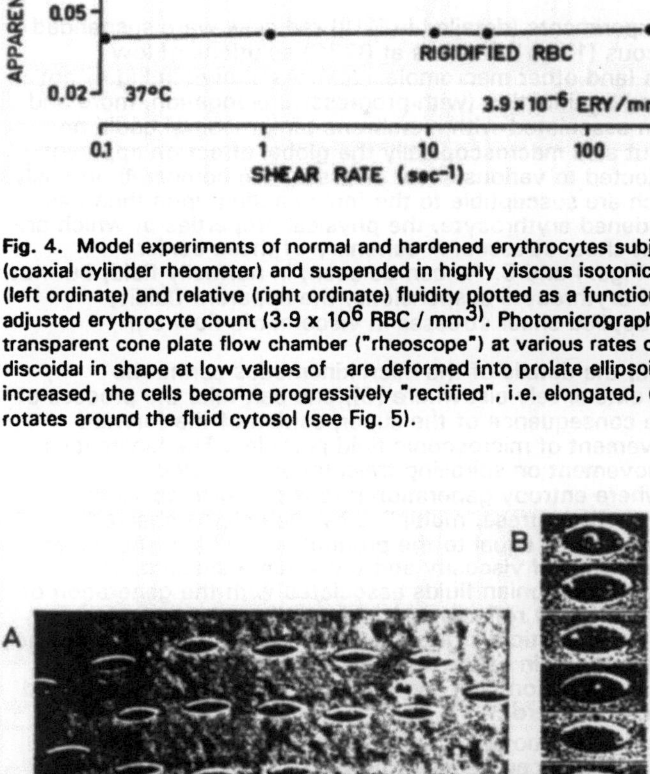

Fig. 5. A. Stationary orientation of rectified erythrocytes ("school of fish") in viscometric flow (230/s). B: Attached latex particles demonstrate membrane rotation while entire cell remains in stationary orientation and elongation.

established fact that the viscous energy dissipation rate associated with Poiseuille-flow of the plasma alone (which contains small amounts of dissolved oxygen) is increased only by a factor 1.1 to 2.0 when red cells are moving along with the plasma. The latter, however, chemically bind roughly 100 times more oxygen than plasma does due to physical solution excess entropy generation falls with the induction of "rectified flow" of the fluid-drop like red cells.

When at rest (i.e.near or at fluid-dynamic equilibrium), the erythrocytes behave like quasi-solid, when rapidly moving they behave like quasi fluid particles. In artificial systems (human red cells suspended in highly viscous dextran solutions) and in viscometric flow (transparent plate viscometer) the transition can be phenomenologically observed and dynamically measured (see Figs. 4 and 5).

In simple viscometric experiments (detailed in [20]) red cells were suspended in isotonic, but highly viscous (12 to 60 mPa s at 37°C) solutions of low molecular weight dextrans (and other macromolecules). As shown in Fig.4, not only the local shear induced rectification (with progressive elongation, more and more stationary orientation associated with membrane tanktreading) could be studied microscopically, but also macroscopically the global effect on apparent fluidity of suspension subjected to various shear stresses. We compared normally flexible erythrocytes, which are susceptible to the forces acting upon them, as well as experimentally hardened erythrocyte, the physical properties of which are independent of the incident shear stresses. A summary of these earlier experiments are shown in Figs.4 and 5. As can be seen, in normally fluid, but not in experimentally hardened erythrocyte suspensions, the apparent fluidity progressively rises as the applied shear stresses in viscometric flow are increased.

For reasons of space, the details of the microkinematics cannot be elaborated here as shown extensively elsewhere [6], one can take the process of momentum transfer as the consequence of the superposition of the forward movement on thermal movement of microscopic fluid particles. The kinematics can be simplified as the movement on spiralling trajectories (depicted schematically in Fig. 6), where entropy generation rate is proportional to the product of shear rate times, shear stress, multiplied by the volume sheared.

This product can be taken as equal to the product of the local shear rate ($\dot{\gamma}^2$), multiplied by the coefficient of viscosity and the volume sheared. The highly ordered movements of Newtonian fluids associated with the generation of the parabolic velocity profile (Fig. 7) reflects a situation where the entropy generation rate shows a characteristic distribution with maximum values near the wall (where the difference in the gain of the spirals depicting the velocity difference is maximum) and the region near the tube axis (where shear rates and thence entropy generation rate is zero).

When fluid erythrocytes are moving along with a newtonian fluid, a very complex flow situation evolves: as can be seen, however, there is a new distribution of velocity gradients: the high apparent fluidity of the cell-plasma

Fig. 7. Schematic representation of the different rates of entropy generation ($ds/dt = f(\dot{\gamma}^2)$) in ▶ fully developed Poiseuille flow depicted as the mutual gliding of fluid lamellae pass each other. Spirals as explained in Fig. 6 of different gain display the low velocity in the layers near the wall the maximum velocity in the layers near the tube axis. However, as there are particles of different forward velocities and thence velocity gradients or rates of shear; the collisions between particles induced by laminar shearing give rise to momentum transfer between faster and slower particles: this process is depicted by spirals of highly different gain. On the shoulder of the parabolic profile, where there is an intermediate gain in individual trajectories the locally different gain is shown by three spirals.

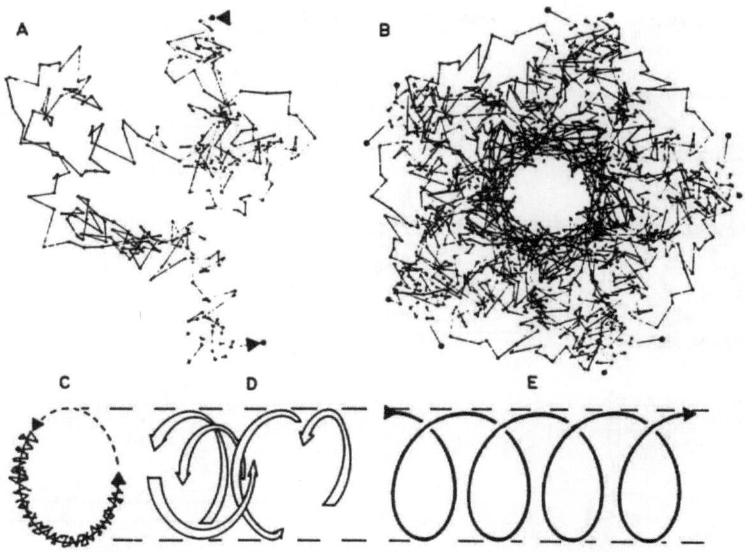

Fig. 6. Schematic representation explaining the superposition of thermal movement and forward movement of fluid elements subjected to convective laminar flow. A: Random walk of one molecule without laminar shear flow. B: Superposition of 5 molecules starting five different places resulting in an net apparent movement along rings. C: Simplified representation of B with superposition random walk upon microscopic circular movement without laminar shear flow. D: Projection of schematized forward movements of trajectories as seen in C after induction of laminar flow. E: Highly simplified trajectory of fluid particle subjected to finite forward velocity

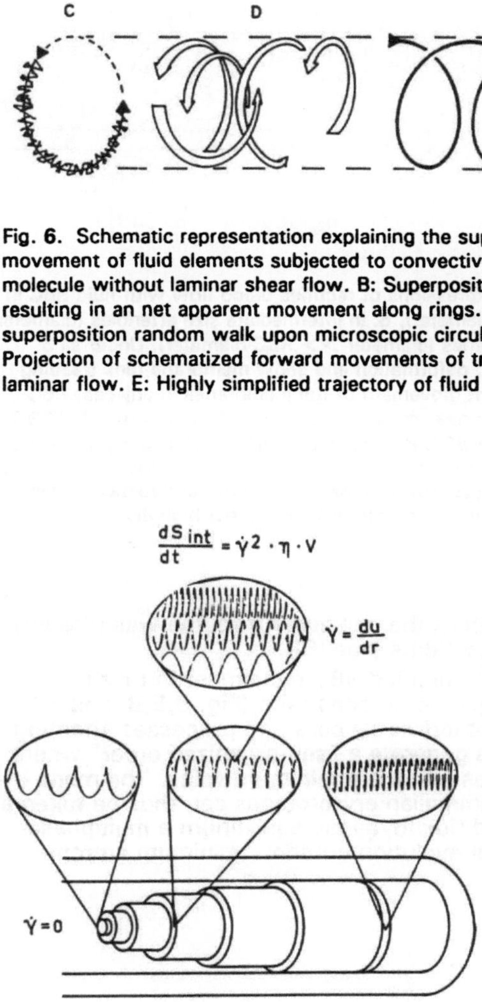

(Fig. 7)

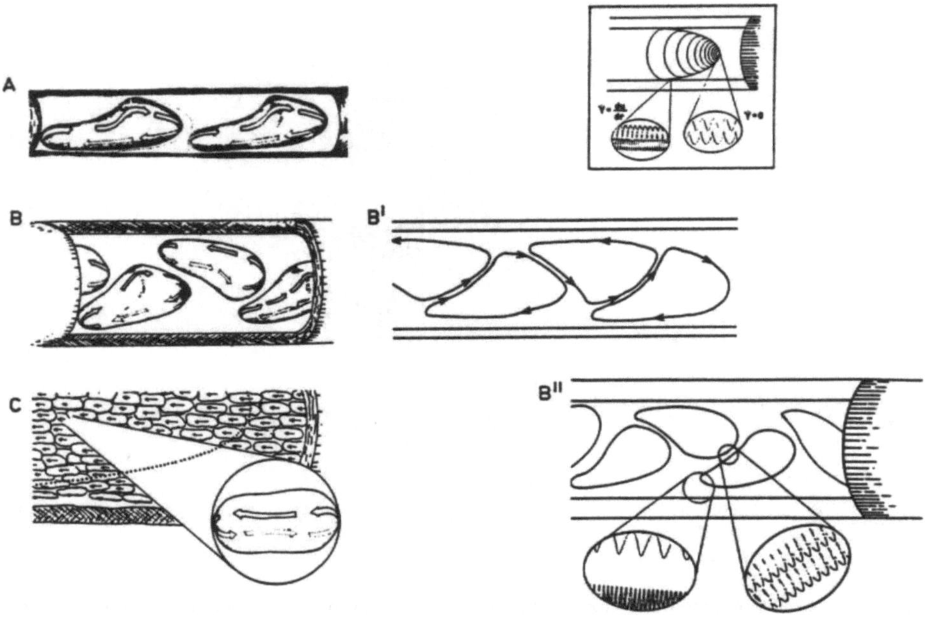

Fig. 8. Schematic representations of various expressions of rectified blood flow with fluid RBC in A: very narrow capillaries (diameter << RBC-diameter), B: in intermediate size arterioles (diameter = RBC diameter at rest), C: in macroscopic arteries (diameter >> RBC-diameter). Under all 3 conditions the cells are in stationary orientation, deformation and show membrane tank treading (= "RBC-rectification"). B': the globally coherent movement of the membranes in adjacent RBC and the formation of a marginal lubricating layer near the wall is shown in "globally rectified RBC flow" (termed "zipper flow" of "slipper-shaped RBC". B'': the decrease in local velocity gradients between moving red cell membranes (identical velocity of membrane gliding past each other. Note also decrease in global velocity gradients: both processes amplify each other and reduce global entropy generation rates in the marginal region (high velocity differences but high distance between cells and tube ball).

mixture being related to the passive drift of the cell away from the wall (thereby reducing the local and global velocity gradients (see [6]).

As depicted for the case of highly rectified RBC movement in tubes (diameter 8-10 μm) where there is no geometric constraint (Fig. 8,B,B' and B'') processes operating in the movement of individual cells and processes affecting the entire ensemble of cells and plasma generate a "self-organized order" where local and global entropy generation rates assume a minimum value. The mode of operation of "dynamically rectified" mammalian erythrocytes can thus be taken as an example where far from thermo- and fluid dynamic equilibrium a multiphase system becomes ordered by the general evolution criterion (minimum entropy principle).

2.3 Synergetic Aspects of Efficacy and Complexity in Interconnected Systems

So far, we have dealt with a very simple and primarily biomechanical phenomenon, which amplifies blood flow. Biological systems are, of course, closely interconnected and exposed to a multitude of flows of energy and matter of highly differing magnitude. Very obviously, the global efficacy is determined by the energetic expenditure of those flows of energy and matter with the greatest power.

On the other hand, the fine tuning of the many different other, often very discreet and low powered flows of energy and matter are as important for the overall biological performance on a local and a global level, on short, but especially on all longer time scales. Here, not just nutritional, but also the many inflammatory responses and reactions maintaining thousands of chemical "flow" equilibria must be considered. It appears that both the high complexity and the remarkable adaptability of compound biological systems are therefore maintained practically free of extra energetic expenditure.

Moreover, as the convective and the diffusive transport rates in the extracellular spaces, and the active and the passive transport systems at the boundary between the intracellular and the extracellular are all limited by the oxygen supply. The synergetic system of rectified erythrocyte transport proves to be pivotal for the "homeostasis" or, in the present semantics - the maintenance of the "boosted steady state" in a l l other systems listed in Table 1.

The power in the flows of ions, of interstitial fluid, of protons, of metabolites and catabolites and of mediators in the blood stream and in the interstitial states can be estimated to amount to less than 1 % of the power associated with the convective movement of the blood, which, of course, is the dominant process dissipating the mechanical energy converted in the myocardial cells. For all these flows, the movements can be said to be "synergistically ordered", meaning that - with respect to the maintenance of the boosted steady state - the sum of the flows is more than the indiviual flows would achieve. Thermodynamically speaking, all these flows can be assumed to be governed by simple principles of minimum entropy generation. Regarding the convective movements of fluids and matter dissolved in them, one can say that they are being enslaved by the dominant power generated in the flow in the systemic circulation (a matter that will be dealt with elsewhere [6]).

Before entering into a discussion of the normal and the disturbed microvascular systems, one additional, namely a structural fact must be mentioned in passing: the geometrical as well as the topological features of the different microvascular beds is dynamically adapted to the physiological function, most probable being fine tuned to the nutritional needs under maximum load. As one of the most important boundary conditions for the operation of microvascular networks, one must keep in mind that for the erythrocytes must be dynamically rectified in order to negotiate the nutritive capillaries with diameters (5 μm) far below that of the resting erythrocytes (7.5 μm) (see Figs. 3A and 8A). Thus, it is not exaggerated to state that the described form of erythrocyte rectification, along with the emerging synergetic ordering and amplified flow due a principle of minimum expenditure acting far away from fluid-dynamic equilibrium is conditio sine qua non for the operation of the entire cardiovascular-pulmonary system.

It is not easy to find an appropriate example from other domains of dynamic systems to illustrate the essentiality of erythrocyte rectification and the resulting reduction in entropy generation rates for microvascular flow: perhaps one might compare it to the essential induction of extreme cooling and thence extreme increase in electrical conduction for the proper operation of magnets used to induce nuclear magnetic resonance in tissues or chemical samples. In NMR systems and in microvascular beds, the boundary conditions for the induction of the flow of electrons and erythrocytes, respectively, are chosen in such a fashion that the entropy generation in the principle channels (due to the induction of incoherent movements) is reduced "at all cost"[3] .

Be that as it may: the following analysis of a much more comprehensive system and of the multiple reactions taking place in them starts from the assumption that the induction of functional efficacy by synergetically ordered mode of operation in the principal conductor of energy and matter induces stability of operation since the flow in the functionally preferred mode very likely enslaves all other contingent flows as well.

3. The Self-Organized Synergetic Order in Microvascular Beds is Disclosed by Disappearance in "Low Flow States"

3.1 The Normokinetic State

All aspects of "synergetic rectification" must be fullfilled in order to allow coherent blood flow in micro vascular networks. During the progress from phenomelogical to the dynamic analysis of the process of blood movement through the terminal microvascular bed, it soon became evident that extremely steep pressure gradients normally act along the arterioles and capillaries. Moreover, it became evident via topological and hydrodynamic network analysis (Schmid-Schönbein [24]) that the contractile state of arterioles (termed "conductance regulators" by us, conventionally classified as "resistance vessels") is the sole "ordering agency" (see below) not just for bulk volumetric flow, but for transmural fluid exchange. The latter is based on filtration (or reversed osmosis) at the arteriolar end, and on reabsorption (or "colloid-osmosis") at the venous end of exchange capillaries. Furthermore, the regulated "attendance" of a greater or lesser number of red cells to exchange capillaries (a phenomenon previously called capillary recruitment and derecruitment) and thence both the convective as well as the diffusive supply (and the diffusive clearance) of metabolites to and from the parenchymal cells (see Fig.9) is ordered by the contractile state of the arterioles.

While admittedly the intravascular, transmural and interstitial "hematokinetics" (defined as the more or less coherent movement of blood constituents) varies widely in different tissues, one can already state that the *normal* state is characterized by almost complete biophysical and biochemical coherence of the movements taking place, while the *pathological* state is characterized by a more-or-less complete *breakdown* of this self-organized order. The latter is replaced by *incoherent* hemodynamic, fluid dynamic and enzymatic *reactions* known to accompany all forms of *low flow states* (or the *hypokinetic circulation*). In this context, we have learned to appreciate that the normally perfused microvascular bed represents a typical example of a system driven to a "boosted steady state" exhibiting "synergetic order"under normokinetic conditions. In all hypokinetic states (see [9])"infrasynergetic" or "microscopic chaos" sets in which is fundamentally different from the "suprasynergetic

---

3 It goes without saying that this analogy only relates to the e x c e s s entropy generation caused by the superposition of erythrocyte movements over that of cell free plasma. However, if one makes a true energy balance, the overall increase in the efficacy is comparable in both systems (if not superior in the erythrocyte movement) since obviously much entropy must be generated to produce the very low absolute temperatures necessary to induce superconduction in the part of the system conducting the bulk of the power.

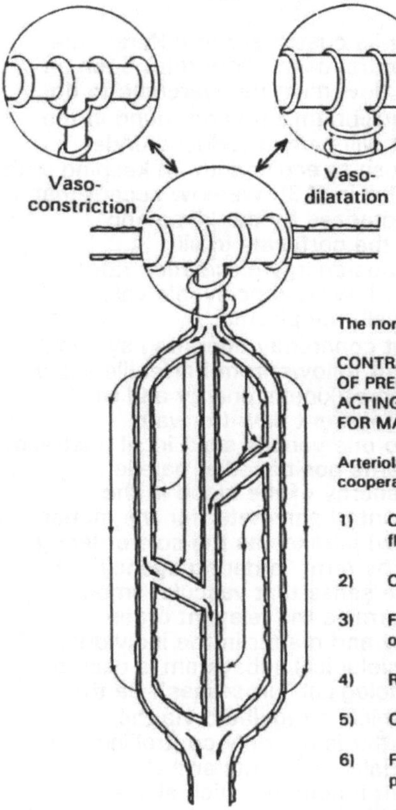

Fig.9. Schematic representation of the mode of operation of normal microvascular networks perfused with "rectified red cell suspensions". The perfusion of microvascular modules and the percolation of the interstitial spaces is exclusively regulated by the contractile state of the precapillary arterioles (details see text), the entropy generation being near its theoretical minimum (namely that associated with perfusion of cell free plasma). Thence, the conductance of the precapillary arterioles limits local pressure gradients, flow rates and functional coherence of intravascular and transmural fluid movements in the "normokinetic state".

chaos"[4] (or deterministic chaos in the sense of conventional non-linear, non-equilibrium thermodynamics).

---

4 The latter can also be found under exceptional conditions of the circulatory systems, e.g. exhausting physical exercise, states of several neurological (spinal injury), the latter not being compatible with life except with the help of contemporary intensive care medicine, see [8].

3.2 Establishment of the "Boosted Steady State" Produces the Non-equilibrium, Manifestation of "Rectified Flow" causes Non-linear Reactions

In phenomenological circulatory physiology, it was customary to differentiate between the "vasomotor" and the "viscous" control of the flow rate. Such an idea of "competition" between determinants of flow must be overcome to do justice to the dynamics prevailing in the non-equilibrium situation found in the normally perfused microcirculatory networks. It will be reintroduced under different aspects when discussing situations close to equilibrium. In keeping with the facts mentioned above and described in detail in [13], we now accept that the *normokinetic state* of the circulation is characterized by the kinematic situation where *erythrocyte rectification* fluidizes the perfusate in all microvascular segments, thereby reducing viscous energy dissipation rates in concentrated red cell suspensions down to their lowest conceivable value, namely that associated with the movement of cell-free plasma.

The global network is a highly complex yet coherently operating system, driven far away from fluid-dynamic equilibrium. It follows from Poiseuille's law and Kirchhoff's rules that the viscous dissipation of kinetic energy and of potential energy in any one segment produces coherent flow (forward displacement of fluid from one arterial source to one venous sink) in all upstream and downstream segments, as well in all segments positioned in parallel.

In this situation, the kinetic and potential energy of the blood in the macroscopic arterial blood vessels is the sole control parameter for the motion, while the fluctuating arteriolar "tone" (see below) is taken as the sole ordering agency for the transfer of energy and matter. The term "ordering agency" (German "ordnende Instanz") is used here in the sense that vascular smooth muscle cells possess a unique capability to determine the relevant order parameters for various coherent flows of energy and matter in the individual segments of microvascular networks. This physiological subsystem is used to instrumentalize the influence of various other biological "processes" via their hydrodynamic effects on input conductance, which is regulated via the contractile state of smooth muscle cells. The latter is not only controlling the overall hydraulic conductance through the arteriolar segments and all downstreams, but also a host of flow-dependent reactions which all are "enslaved" by one control parameter and one ordering agency producing a wide spectrum of functionally coherent biological reactions ("synergetic order"). As will be discussed below, the infrasynergetic chaos in the disturbed, hypokinetic circulation is characterized by the operation of a host of different local ordering agencies and thence by global incoherence.

In the normokinetic situation, however, the biological activity called "tone" (or contractile state) of the smooth muscle cells is highly variable, owing to its *susceptibility* to a host of *biophysical, biomechanical, metabolic, local chemical, nervous* and *systemic hormonal influences*, all capable of exerting the roles of transient "attractors" for the dynamic play of contraction and relaxation of the vascular smooth muscle. Consequently, since the controlling agency (and thence the hydraulic conductance of the flow controlling arterioles) fluctuates in response to the variable action of many different biological processes taking place within the vessels, in the extravascular so do all processes taking place and of course in the intracellular space of the parenchymal cells supplied by a vascular network, a topic extensively discussed in [6].

3.3 The Hypokinetic State Operating near Equilibirum

In all cases where the self-organized stabilization of the circulation under conditions of physical rest breaks down, so does functional coherence. This can

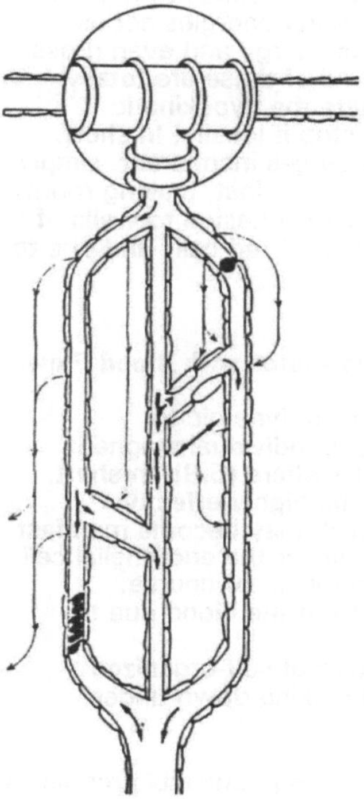

Fig. 10. Destabilized or "hypokinetic" microcirculation associated with arteriolar relaxation but locally curtailed conductance due to intravascular obstacles becoming functional in the absence of the physiologically high shear stresses. Multiple ordering agencies destroy coherent perfusion (details see text)

occur in all hypokinetic situations (called "low flow states") such as those that prevail due to cardiac pump failure or due to uncompensated hypovolemia, due to macrovascular obstruction at the level of arteries and veins or due to a host of inflammatory, metabolic, immunological, and toxic disturbances affecting the parenchymal cells and/or the microvessels. Interestingly, such loss of microvascular coherence is also found in an important "molecular disease", namely sickle cell anemia, which is caused by erythrocyte rigidification [20]. All these hypokinetic situations are characterized by the complete lack of coherence of all biophysical, but soon also of the biochemical reactions taking place in the intravascular, the interstitial and the intracellular space (see [9]) Initially reversible, later irreversible combinations of malfunctions occur, highlighted by the progressive escape of the many different energy conversion processes from the one global ordering agency controlling the boosted steady states of the healthy system. The contractile tone of arterioles (and thence its susceptibility to regulating processes) is progressively lost, while at the same time new "strange attractors"[5] assume the role of local ordering agencies [9]. Suffice it here to direct the attention of the reader to an extensive, but by no means comprehensive list of local "controlling reactions" exerting influences in the disturbed microcirculation (Fig.10).

Note that these act exclusively when the dominating influence of rapid flow (high local kinetic energy) is no longer acting, so that weaker energies act as attractors. These include adhesive energies, gravitational energy and even gross geometrical abnormalities (due to endothelial swelling). All of these are totally irrelevant under normokinetic conditions but can dominate the hypokinetic situation where they not just retard flow, but can even stop it locally. In short, the flow within the different microvascular segments becomes incoherent, simply because the high mobility of dynamically fluidized perfusate is lost, making room for a shear-dependent local viscidation due to plugging and adhesion to walls of leucocytes, due to aggregation and functional rigidification of red cells and due to various thromboplastic reactions (see below).

3.4 Coherent and Incoherent Biological Reactions Associated with Blood Flow

This concept leads to the essential transgression from fluid-dynamic to biochemical, enzymatic and cellular events taking place in individual stagnant microvascular segments: a topic extensively covered elsewhere [5,9]. In short, coherent microvascular flow protects the system from the highly effective procoagulatory potentials of the blood components, which may become manifest when in contact with the subendothelial structures whenever the endothelial cell lining of the microvacular becomes damaged. This situation is, of course, highlighted by a first-order phase transition: solidification of the blood due to coagulation.

Three mechanisms must be mentioned in the context of self-organized antithrombotic reactions[6] taking place under normal, breaking down under pathological situations.

1) The endothelial cells - presumably on the basis of strong coulombic repulsion normally preclude the close physical contact and thence the accumulation of macromolecules with the solid elements of the wall that can occur after pathological sloughing of endothelial cells.
2) The blood stream automatically produces "convective dilution" of any procoagulatory agent, especially of autocatalytically activated coagulation enzymes: this breaks down after local stagnation of flow.
3) The blood stream automatically enforces the reaction between coagulation enzymes and the "inhibins" and "scavengers". Pars pro toto the reaction of antithrombin III (binding not only thrombin but other members of the coagulation cascade), and of albumin (binding arachidonic acid liberated from injured cell membranes) are mentioned. Kinematically speaking, the coherent movement of the blood connects any possible site of injury-induced generation of thrombin or arachidonic acid with an infinite reservoir of "inactivating" material forming complexes that are removed into an infinitely large sink for the inactive complexes (Antithrombin-Thrombin, Albumin-Arachidonic acid). All of these reactions are locally prevented in these microvascular segments where blood flow goes to zero.

4. Generalizations Defining Synergetic Modes of Operation in Physiology

The operation of the vasoconstrictor controlled microvascular networks at rest serving as the "conductances" linking the macroscopic arterial and the macroscopic venous systems can be said to be kept not only in the "boosted steady state", but the arterioles can said to be "tensilized" due to pressure-induced vasoconstriction. This limits the export of energy and matter, the storing energy and matter upstream of dynamically "tensilized" macromolecular structure. Accelerated export of energy and matter from the arterial system (and thence coherent perfusion of the microvascular network as discussed in Section 3.1.) can be initiated by a mere relaxation of the permanently induced active tension in the contractile elements of the arteriolar walls. Not only is the energy of the "stimuli" extremely small in comparison to the effects elicited, but the increment in flow for any given pressure gradient is favoured by the highly non-linear consequences of Poiseuille's law [$Q = f(r^4)$]. As the diameter of the "bottle necks" of the microcirculation are smaller than the resting diameters of the vast majority of the cells negotiating them, the shear-induced "erythrocyte rectification" further amplifies flow (leading to flow acceleration favoured by a fall in excess entropy generation after erythrocyte rectification).

It would appear as if this well-studied mesoscopic example of biological self-organization is typical for the synergetic mode of operation as it can be assumed to occur e.g. in neurons and neuronal chains and in other physiological functions (a topic to be discussed elsewhere [6]).

5. Epilogue

The concepts developed and the semantics used in statistical mechanics, in non-equilibrium thermodynamics and in synergetics not only prove to be extremely helpful in comprehending complex biological systems, but they provide a logical framework allowing us to put so-called anomalies (e.g. highly non-linear pressure flow relationship causing the non-linear anomalous viscosity of blood as a multiphase fluid) into harmony with non-linear system dynamics developed in other fields. For this purpose, however, a mutual process of "linguistic approximation" (and perhaps even conceptual approximation) must be initiated: one might call for a process of "dissipative structuring" of semantics (following Haken's concepts about synergetics of linguistic social systems). After its successful application, a very efficient razor of Occam will most likely be at our disposal. In the field of expertise of the present author, this idea can be exemplified.

In replacing a babylonian confusion of terms used in rheology (and especially hemorheology) one can delineate obvious, but previously overlooked analogies between the mode of operation of blood elements in microvascular networks and other highly self-organized physical processes paradigmatic for spontaneous ordering occurring exclusively when matter is driven under steep potential gradients. The induced order is automatically lost whenever control parameters fall below critical thresholds. It is by no means redundant to stress again that it was such an event, i.e. the observation of a special form of "critical slowing" that brought to our attention that the normal function is not only spontaneously ordered, but might actually be operating in "synergetic cooperativity".

Critical experiments providing the quantitative proof for the assumed synergetic order are missing at present: there is, however, an abundance of indirect evidence in the biomedical literature that can be used to this end. As shown elsewhere [13], a decrease in excess entropy generation (as a proof of the suspected synergetic mode of operation) in microvascular blood flow can be easily obtained by calculating appropriate functional order parameters (for example the amount of energy transferred, normalized by the amount of energy dissipated). All forms of "hypokinetic microvascular disturbances" (conventionally called "low flow states") show a dramatic drop in the efficiency, an experimental fact that in the future can be expressed by a fall in the numerical values of appropriate order parameters that remain to be developed.

It is also important to stress again in this context that the actual reason for an induced low flow state is often quite irrelevant, they all share the features of incoherent microvascular blood flow described in sections 3.2 and 3.3, they are all associated with often dramatic decrease in local oxygen supply without commensurate decrease in global blood flow [9]. Incidentally there is a substantial p r a c t i c a l significance associated with the application of synergetic concepts, namely that related to therapy of hypokinetic circulatory states [25]. We have convincing evidence that procedures improving the fluidity of the blood are very successful remedies for all kinds of "low flow states": the best and simplest explanation for this is that the normokinetic, self-organized mode of operation is being restored [9].

References

1. H. Haken: This volume
2. H. Haken und Maria Haken-Krell: *Entstehung von biologischer Information und Ordnung* (Wissenschaftliche Buchgesellschaft, Darmstadt 1989)
3. H.P. Koepchen: Physiology of Rhythms and Control Systems: An integrative Approach, in: *Rhythms in Physiological Systems* (Springer Berlin, Heidelberg 1991), 3-20
4. L. von Bertalanffy: *Biophysik des Fließgleichgewichts. Einführung in die Physik offener Systeme* (Vieweg, Braunschweig 1953)
5. H. Schmid-Schönbein: Synergetics of Fluid-Dynamic and Bochemical Catastrophe Reactions in Coronary Artery Thrombosis, In: *Unstable Angina* (eds. W. Bleifeld et al.), (Springer, Berlin, Heidelberg 1990), pp. 16-51
6. H. Schmid-Schönbein: Einführung in die physiologische Synergetik, in: *Int.Textbooks of Physiology* (eds. K. Koepchen und Mommertz), (Springer, Heidelberg in press)
7. B. Hille: *Ionic channels of excitable membranes*. Chapter 10 Selective Permeability: Independence. (Sinauer, Sunderland Mass. 1984)
8. H. Schmid-Schönbein and S. Ziege: The high pressure system of the mammalian circulation as a dynamic self-organization system, in: *Rhythm in Physiological Systems* (eds. H. Haken, H.P. Koepchen), (Springer Berlin, Heidelberg, New York 1991), 77
9. H. Schmid-Schönbein: Synergetic order and chaotic malfunctions of the circulatory systems in "multiorgan failure": breakdown of cooperativity of hemodynamic functions as cause acute microvascular pathologies, in: *Update in Intensive Care and Emergency Medicine*, (ed. J.L. Vincent), (Springer Berlin, Heidelberg, New York, London, Paris, Tokyo, Hong Kong 1990, Vol. 11), 3
10. W. Ebeling: *Chaos - Ordnung - Information* (Harri Deutsch, Frankfurt 1989)
11. W. Ebeling and Yu.L. Klimontovich: *Self-Organization and Turbulence in Liquids*. (Teubner, Berlin 1984)
12. H. Haken und A. Wunderlin: *Die Selbststrukturierung der Materie*. Synergetik in der unbelebten Natur (Viehweg, Braunschweig 1991)
13. H. Schmid-Schönbein: Synergetics of O2-Transport in the mammalian microcirculation: Cooperativity of molecular,cellular, kinematic and hemodynamic factors in "rectified blood flow", in: *Biomedical Transport Processes* (eds. F. Mosora, C. Caro, Ch. Baquey, H. Schmid-Schönbein, R. Pelissier, E. Krause), (Plenum Press, New York 1990), 185
14. H. Schmid-Schönbein: *Rheologische Eigenschaften von Erythrozyten*. Habilitationsschrift Ludwig Maximilians-Universität München 1970
15. H. Schmid-Schönbein: Microrheology of Erythrocytes and Thrombocytes. Blood Viscosity and the Distribution of Blood Flow in the Microcirculation, in: *Handbuch der allgemeinen Pathologie III/7 Mikrozirkulation* (ed. H. Meessen), (Springer Berlin, Heidelberg, New York 1977), 290
16. H. Schmid-Schönbein: Blood rheology and oxygen conductance from the alveoli to the mitochondria, in: *Drugs and the Delivery of Oxygen to Tissues* (ed. J.St. Fleming), (CRC Press Boca Raton Fl. 1990), 15
17. J.R. Pappenheimer: Contributions to Microvascular Research of Jean Leonard Marie Poiseuille, in: *Handbook of Physiology* (Amer.Physiol.Society, Bethesda MD 1984), Section 2, vol. IV, Part 1, 1

18. P. Gaehtgens: in: ***Pathophysiologie des Menschen***, (ed. K. Hierholzer und R.F. Schmidt) (Verlag Chemie, Weinheim 1991) 11
19. R.Fahraeus and T. Lindquist: The viscosity of blood in narrow capillary tubes, Amer. J. Physiol. **96**, 562 (1931)
20. H. Schmid-Schönbein: Fluid dynamics and hemorheology in vivo: the interactions of hemodynamic parametrs and hemorheological "properties" in determining the flow behavior of blood in microvascular networks, in: ***Clinical Blood Rheology*** (ed. G.D.O. Lowe), (CRC-Press, Boca Raton, Fl. 1988), p. 129
21. S. Chandrasekhar: Hydrodynamic and Hydromagnetic Stability (Clarendon Press, Oxford 1961)
22. R. Grebe and H. Schmid-Schönbein: Closed Fluid Quadrilamina Model of the Erythrocyte Membrane, in: ***Biomedical Transport Processes***, (eds. F. Mosora, C. Caro, ch. Baquey, H. Schmid-Schönbein, R. Pelissier, E. Krause), (Plenum Press, New York 1990) 223
23. H. Schmid-Schönbein: Blood rheology and oxygen transport to tissue, in: ***Oxygen Transport to Tissue*** (eds. A.G.B. Kovach, E. Dora, M. Kessler, I.A. Silver), (Pergamon Press Akademiai Kiado Budapest 1981), p. 279
24. H. Schmid-Schönbein and H. Goslinga: Review Article: The present state of hemodilutional therapies: potentials and limits of the most potent tool of "clinical hemorheology". Clinical Hemorheology **12(6)**, 883 (1992)

Index of Contributors

Springer-Verlag and the Environment

We at Springer-Verlag firmly believe that an international science publisher has a special obligation to the environment, and our corporate policies consistently reflect this conviction.

We also expect our business partners – paper mills, printers, packaging manufacturers, etc. – to commit themselves to using environmentally friendly materials and production processes.

The paper in this book is made from low- or no-chlorine pulp and is acid free, in conformance with international standards for paper permanency.

MIX
Papier aus verantwortungsvollen Quellen
Paper from responsible sources
FSC® C105338

If you have any concerns about our products,
you can contact us on
ProductSafety@springernature.com

In case Publisher is established outside the EU,
the EU authorized representative is:
Springer Nature Customer Service Center GmbH
Europaplatz 3, 69115 Heidelberg, Germany

Printed by Libri Plureos GmbH
in Hamburg, Germany